W9-AEH-868

ANIMAL AND PLANT

Anatomy

VOLUME CONSULTANTS

• Amy-Jane Beer, *Natural history writer and consultant* • John Gittleman, *University of Virginia, VA* • Tom Jenner, *Academia Británica Cuscatleca, El Salvador* • Chris Mattison, *Natural history writer and researcher* • Graham Mitchell, *King's College, London* • Richard Mooi, *California Academy of Sciences, San Francisco, CA* • Kieran Pitts, *Bristol University, England* • Adrian Seymour, *Bristol University, England*

5

Hummingbird – Manatee

Marshall Cavendish
Reference
New York

Contributors

Roger Avery; Richard Beatty; Amy-Jane Beer; Erica Bower; Trevor Day; Erin Dolan; Bridget Giles; Natalie Goldstein; Tim Harris; Christer Hogstrand; Rob Houston; John Jackson; Tom Jackson; James Martin; Chris Mattison; Katie Parsons; Ray Perrins; Kieran Pitts; Adrian Seymour; Steven Swaby; John Woodward.

Consultants

Barbara Abraham, Hampton University, VA; Glen Alm, University of Guelph, Ontario, Canada; Roger Avery, Bristol University, England; Amy-Jane Beer, University of London, England; Deborah Bodolus, East Stroudsburg University, PA; Allan Bornstein, Southeast Missouri State University, MO; Erica Bower, University of London, England; John Cline, University of Guelph, Ontario, Canada; Trevor Day, University of Bath, England; John Friel, Cornell University, NY; Valerius Geist, University of Calgary, Alberta, Canada; John Gittleman, University of Virginia, VA; Tom Jenner, Academia Británica Cuscatleca, El Salvador; Bill Kleindl, University of Washington, Seattle, WA; Thomas Kunz, Boston University, MA; Alan Leonard, Florida Institute of Technology, FL; Sally-Anne Mahoney, Bristol University, England; Chris Mattison; Andrew Methven, Eastern Illinois University, IL; Graham Mitchell, King's College, London, England; Richard Mooi, California Academy of Sciences, San Francisco, CA; Ray Perrins, Bristol University, England; Kieran Pitts, Bristol University, England; Adrian Seymour, Bristol University, England; David Spooner, University of Wisconsin, WI; John Stewart, Natural History Museum, London, England; Erik Terdal, Northeastern State University, Broken Arrow, OK; Phil Whitfield, King's College, University of London, England.

Marshall Cavendish
99 White Plains Road
Tarrytown, NY 10591–9001

www.marshallcavendish.us

Library of Congress Cataloging-in-Publication Data
Animal and plant anatomy.
p. cm.
ISBN-13: 978-0-7614-7662-7 (set: alk. paper)
ISBN-10: 0-7614-7662-8 (set: alk. paper)
ISBN-13: 978-0-7614-7668-9 (vol. 5)
ISBN-10: 0-7614-7668-7 (vol. 5)
1. Anatomy. 2. Plant anatomy. I. Marshall Cavendish Corporation. II.
Title.

QL805.A55 2006
571.3--dc22

2005053193

Printed in China
09 08 07 06 1 2 3 4 5

Marshall Cavendish
Editor: Joyce Tavolacci
Editorial Director: Paul Bernabeo
Production Manager: Mike Esposito

The Brown Reference Group plc
Project Editor: Tim Harris
Deputy Editor: Paul Thompson
Subeditors: Jolyon Goddard, Amy-Jane Beer, Susan Watts
Designers: Bob Burroughs, Stefan Morris
Picture Researchers: Susy Forbes, Laila Torsun
Indexer: Kay Ollerenshaw
Illustrators: The Art Agency, Mick Loates, Michael Woods
Managing Editor: Bridget Giles

Photographic credits, volume 5

Cover illustration is a composite of anatomical illustrations found within this encyclopedia.

Corbis: Bettmann 617; Kulka, Matthias 614; Lepp, George D. 584; Rotman, Jeremy L. 691; Digital Stock 593, 597, 649, 678; **FLPA:** Lanting, Frans 671; **Natural Visions:** 647; NOAA: 637; **NaturePL:** Blackwell, Peter 601; Foott, Jeff 591; Freund, Jurgen 638, 714; Hall, David 689; Oxford, Pete 581; Perrine, Doug 712; Pusser, Tod 716; Ruiz, Jose B. 633; Shah, Anup 599, 605; Shale, David 643; Stammer, Sinclair 645; Vezo, Tom 583; Watts, David 653; **OSF:** Gutierrez Acha, Joaquin 634; Haring, David 625; Macewen, Mark 628; Phototake Inc. 607bl, 607br, 607cl, 607cr, 607tl, 607tr; **Photodisc:** 669, 675; **Photos.com:** 587, 663, 665, 680, 683; **Rex:** Voisin/Phanie 620, 623; **SPL:** Bsip Lecaque 616; McCarthy, David 609; Menzel, Peter 622; Scharf, David 693; Steger, Volker 699; **Still Pictures:** Aitken, Kelvin 640, 685; Bavendam, Fred 704, 707; Cristofori, Marco 682; Faulkner, Douglas 710; Harvey, Martin 627; Henno, Robert 701; Murawski, Darlyne A. 698; Reschke, Ed 695; Seitre, Roland 652, 660; Watts, David 658.

Contents

Hummingbird

ORDER: Trochiliformes FAMILY: Trochilidae
GENERA: *Archilochus* and around 108 others

Hummingbirds are small, mostly nectivarous (nectar-eating) birds that are found exclusively in the Americas. They live in a wide variety of habitats, from suburban backyards to tropical rain forests and mountain grasslands. There are about 335 species of hummingbirds in 109 genera, and the greatest diversity lives in montane forests in northern regions of South America.

Anatomy and taxonomy

Scientists organize organisms into taxonomic groups based largely on anatomical features. Hummingbirds have their own taxonomic group—the family Trochilidae. Their closest relatives are the swifts (family Apodidae), which have a wider distribution.

• **Animals** All animals are multicellular and feed on other organisms. They differ from other multicellular life-forms in their ability to move around (generally using muscles) and to respond rapidly to stimuli.

• **Chordates** At some time in its life cycle a chordate has a stiff, dorsal (back) supporting rod called the notochord.

• **Vertebrates** The notochord of vertebrates develops into a backbone made up of units called vertebrae. Vertebrates have a muscular system consisting primarily of muscles that are bilaterally paired (they lie on each side of a line of symmetry running the lenth of the body).

• **Birds** The second largest vertebrate group, birds evolved from reptilian ancestors more than 150 million years ago. Their most obvious feature is the feathers that cover the body. Birds are bipedal and do not have teeth, and at least some of their skeleton is pneumatized (hollow). Most birds can fly, and all are descended from flying ancestors.

• **Hummingbirds and swifts** These birds are members of the superorder Apodimorphae. All are small birds with 10 primary flight feathers on each wing and 6 or 7 secondary feathers. Most species have 10 tail feathers. All lay white eggs. Young hummingbirds and swifts remain in the nest for a period after they have hatched from their egg (they are nidiculous).

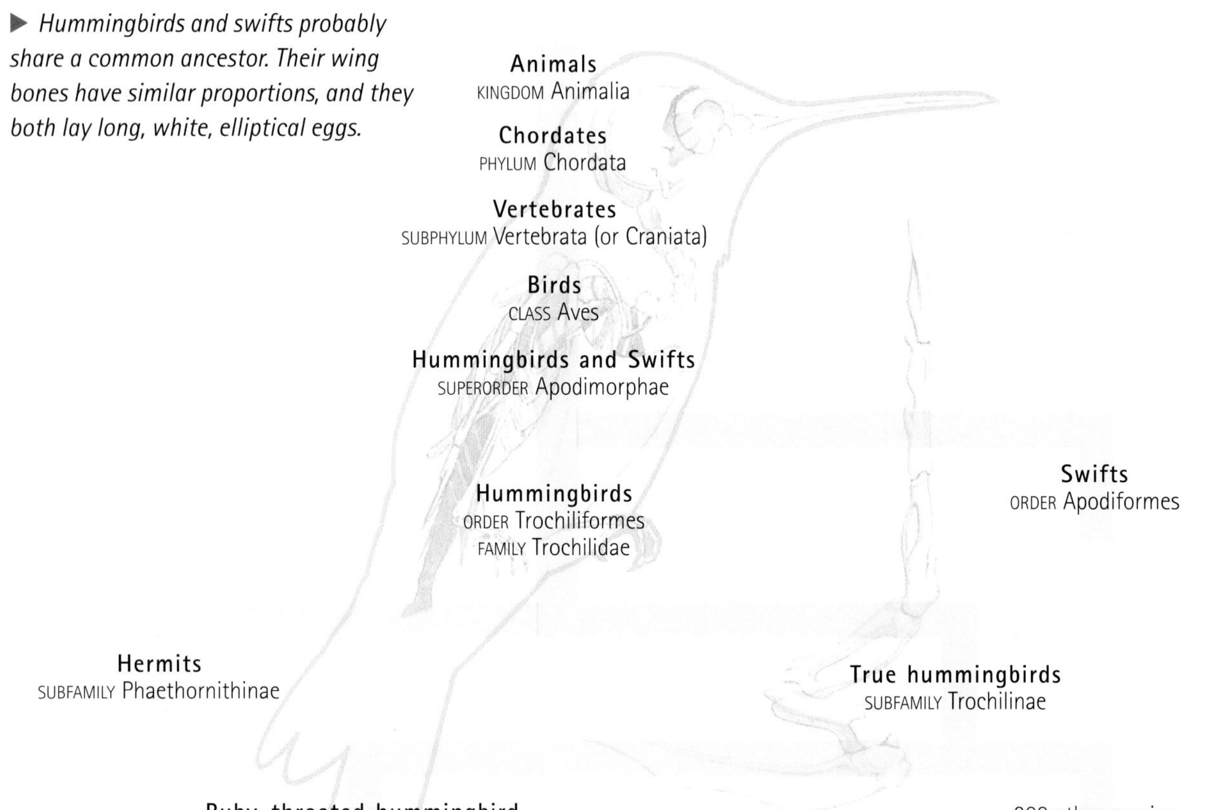

▶ *Hummingbirds and swifts probably share a common ancestor. Their wing bones have similar proportions, and they both lay long, white, elliptical eggs.*

▲ *Hummingbirds spend a great deal of time lapping nectar from flowers, and the shape of their bill suits this diet.*

● **Hummingbirds** These small birds (order Trochiliformes; family Trochilidae) live only in the Americas. They have very iridescent plumage and eat nectar and insects. Hummingbirds have a long, slender bill, small legs, and short arm bones (but long hand bones) in each wing. They are capable of prolonged hovering as well as flying forward and backward. The 34 or so species of hermit hummingbirds (subfamily Phaethornithinae) generally have a long and often curved bill, and insects make up a relatively large proportion of their diet. The 300 or so species of true hummingbirds (subfamily Trochilinae) display a great array of anatomical features, with variation in body size, plumage color, and bill length and shape.

● **Ruby-throated hummingbird** This tiny bird weighs just 0.11 ounce (3.2 g) and has a wingspan of only 4.5 inches (11.5 cm). Both sexes are mostly green on the upperparts, and females have white tips on their tail feathers. Adult males have an iridescent deep red breast.

FEATURED SYSTEMS

EXTERNAL ANATOMY Hummingbirds are small flying birds with relatively long wings, short legs, and iridescent plumage. They display an extraordinary range of bill lengths and shapes. *See pages 582–584.*

SKELETAL SYSTEM A large keel, to which the flight muscles attach, helps hummingbirds beat their wings extremely fast. This feature, coupled with long hand bones in the wings, enables them to hover for prolonged periods and even fly backward. *See page 585.*

MUSCULAR SYSTEM Up to 30 percent of a hummingbird's weight is concentrated in the two major muscles that power the bird's flight. *See page 586.*

NERVOUS SYSTEM A hummingbird's brain is among the largest of any bird's, relative to body size. Hummingbirds have highly developed senses of sight and hearing, but poor senses of smell and touch. *See page 587.*

CIRCULATORY AND RESPIRATORY SYSTEMS Blood is pumped around the body by a heart that is larger, in relation to body size, than that of any other endothermic (warm-blooded) animal. Like other birds, hummingbirds have a respiratory system consisting of air sacs, in addition to their lungs. *See page 588.*

DIGESTIVE AND EXCRETORY SYSTEMS Most species of hummingbirds eat mainly nectar, but a few have a diet of mostly insects. Individual hummingbirds often consume more than half their total weight in food every day, and may drink twice their weight in water. *See pages 589–590.*

REPRODUCTIVE SYSTEM Like other birds, hummingbirds reproduce sexually. Females lay at least one clutch of two pure white eggs every year. Females build the nest, incubate the eggs, and feed the dependent chicks almost entirely without help from male birds. *See page 591.*

External anatomy

CONNECTIONS

COMPARE the weak, tiny legs and feet of a hummingbird with those of an ***EAGLE*** or ***OSTRICH***.

COMPARE the long, fine bill of a hummingbird with that of a ***PENGUIN*** or ***WOODPECKER***.

COMPARE the flying skills of a hummingbird with the flightless ***OSTRICH*** and ***PENGUIN***.

A hummingbird shares the basic features of its external anatomy with other birds: its head and body are covered with feathers; it has two wings and a tail composed largely of feathers; it has two legs, two eyes, and a bill. All hummingbirds are small or very small. Even the biggest species, the giant hummingbird, weighs only about 0.8 ounce (22 g). The bee hummingbird weighs only 0.07 ounce (2 g); it is the smallest endothermic (warm-blooded) vertebrate. The toes of hummingbird feet are capable of closing around a small branch when the bird perches, but the legs are not capable of supporting movement on the ground, so a hummingbird always moves by flying. Feathers provide protection, insulation, the means to fly, and distinctive coloration—this is no different from most birds, but there are probably more feathers per square inch on a hummingbird's body than on any other type of bird.

Some species of hummingbirds have a very long, fine bill, which the bird can insert deep into tube-shaped flowers to feed on nectar. Others have a fine but shorter bill that operates as a feeding tool in smaller flowers. Still others have a stubbier bill that is better suited for catching insect prey. Hummingbirds in the genus *Ramphomicron* have a bill that is less than

▶ **Male ruby-throated hummingbird**
Hummingbirds have remarkable flying skills, but their legs are too short and weak for walking. Characteristic hummingbird features, in addition to their small size, are the long, fine bill and the iridescent plumage, particularly of males.

A long **bill** *enables the bird to probe deep into flowers in search of nectar.*

The green **back feathers** *help camouflage the hummingbird against the leaves of trees.*

The highly iridescent **throat feathers** *identify this individual as an adult male.*

The **covert feathers** *lie on top of, and support, the flight feathers.*

The longest feathers in the wing are called the **primaries**. *They are the main flight feathers.*

Three **toes** *face forward, and a fourth points backward.*

There are five pairs of feathers in the **tail**.

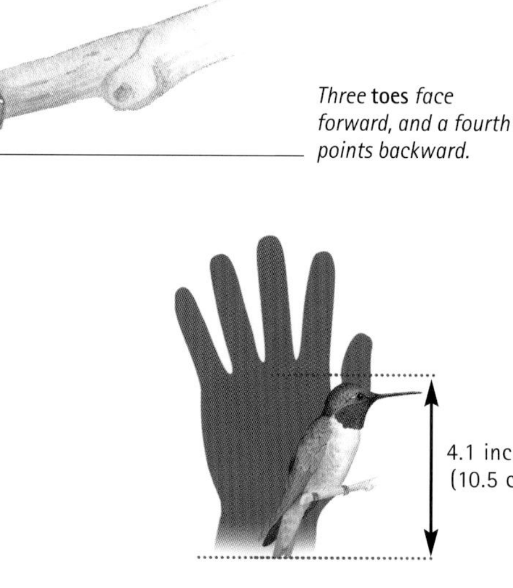

IN FOCUS

Flight

Most birds generate the power to fly during the downstroke of their wings, but hummingbirds are also able to generate power during the upstroke. They can fly backward as well as forward, and they can hover. During forward flight, the tips of the wings move in a vertical oval in the air, as is the case in other birds. When a hummingbird hovers, its wing tips move in a horizontal figure-eight pattern. By tilting the horizontal plane of movement slightly downward, the bird is able to move very slowly forward; and by tilting the plane of movement of its wing tips a little above the horizontal, the hummingbird can fly backward. Changes in the plane of motion of its wings create the humming sound from which the birds get their name.

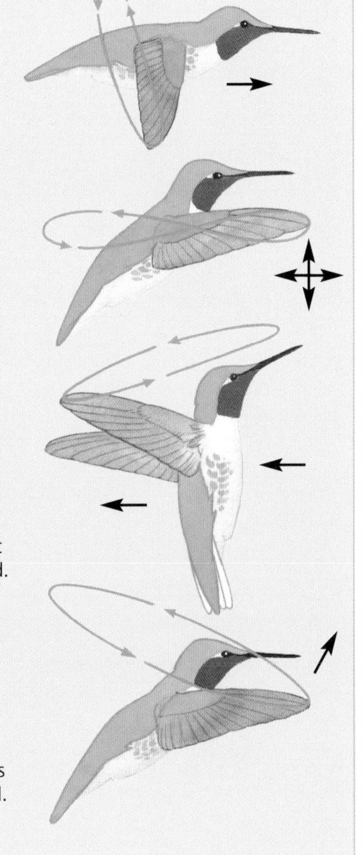

▲ *This black-chinned hummingbird shares with other species an amazing agility in flight. It can also fly great distances, migrating all the way from Mexico to British Columbia in spring.*

0.4 inch (1 cm) long, while some in the genus *Ensifera* are more than 4 inches (10 cm) long. In most species the bill is slightly down-curved, although in a few it is recurved (curved upward), and others have a bill that is laterally compressed (flattened from side to side). The nostrils are covered from above by a shelflike scale, or operculum. This may keep pollen from entering the nostril.

Flying machine

A hummingbird is an amazing flying machine, and this ability is reflected in its wings, which are relatively long in relation to body size. Like any other small bird, a hummingbird needs to be able to beat its wings rapidly to remain airborne. However, the extreme speed of a hummingbird's wings enables it to hover and fly backward as well as forward—very useful skills in maneuvering around flowers. The giant hummingbird averages a very rapid 10 to 15 wing beats per second, the amethyst woodstar beats its wings an astonishing 80 times each second, and other species may reach 200 beats per second during courtship displays. Individual

CLOSE-UP

Feather iridescence

The feathers on the body of a hummingbird are extremely small and closely packed. A female Allen's hummingbird, for example, has more than 1,600 feathers, compared with 1,900 on a brown thrasher, which has a skin surface area about 10 times larger. Apart from the flight and tail feathers, most of a hummingbird's feathers are iridescent, especially those on the breast. In fact, only the distal section of each feather (the part farthest from the body) is iridescent, but since the feathers are so closely packed, the iridescence appears to cover the whole breast.

Most bird colors are produced by selective light absorption and reflection by melanin pigments in the feathers, but this is not how the bright, variable colors of a hummingbird's iridescence are created. Instead, the iridescence depends on an effect called interference coloration, which also produces the colors seen on a soap bubble or an oil film. This involves light passing through a material—in this case, part of a feather—with a refractive index (ability to refract, or bend, light) different from that of air. Some of the light reflected from the back of the material interacts with that reflected from the front. This causes interference, in which certain wavelengths of light are lost, so normal white light is changed into colored light. For the same refractive index, the thickness of the material through which the light passes also affects which wavelengths are lost, and therefore the color of the emerging light.

The iridescent part of the barbules of hummingbird feathers are densely packed with tiny platelets, only 2.5 microns long and even thinner. The platelets contain even tinier air bubbles. The thickness of a platelet, and the amount of air it contains, determines the color of the light reflected—and therefore the color of the iridescence. Also, the angle at which the light strikes the platelets is important, since light has to travel through a greater thickness of platelet if it does not strike at a right angle. This is why hummingbird feathers appear to change color.

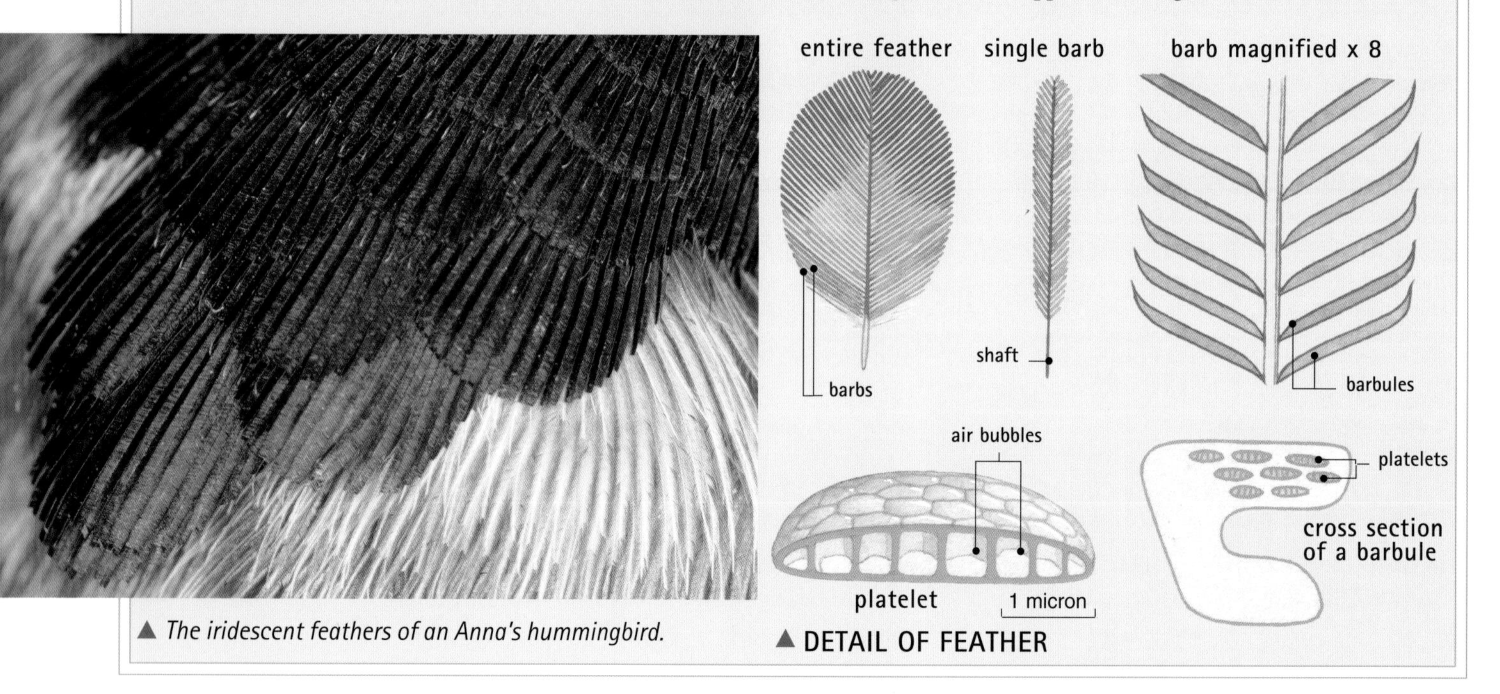

▲ *The iridescent feathers of an Anna's hummingbird.*

▲ DETAIL OF FEATHER

wing beats are invisible to the human eye, and the wings of a flying hummingbird make a humming sound for which the birds are named.

There are 10 primary and 6 or 7 secondary flight feathers on each wing. The innermost primary feather is the shortest, and the outermost is the longest. There are usually 10 tail feathers, although some species have fewer. The shape of the tail varies considerably among species: the tail may be forked, wedge-shaped, rounded, graduated, scissor-shaped, or squared. Although hummingbirds' small size means they do not have as many feathers as much larger birds, they do have a very high feather density. For example, the feathers on a ruby-throated hummingbird are packed together more than five times as densely as those on a brown thrasher.

Skeletal system

CONNECTIONS

COMPARE the deep sternum of a hummingbird with that of a ***PENGUIN***.

COMPARE the short tibiotarsus (digit bone) of a hummingbird with that of an ***EAGLE***.

A hummingbird's basic skeleton is similar to that of other birds. It can be considered in two parts: the axial skeleton, which is made up of the skull, the vertebral column, ribs, and sternum, or breastbone; and the appendicular skeleton, which comprises the wings, legs, and pectoral and pelvic girdles.

However, several parts of a hummingbird's skeleton are distinct from that of other birds. The sternum had a very deep keel (to attach flight muscles); its total depth may be greater than the depth of the bird's rib cage. This provides a large surface area onto which the bird's all-important flight muscles are attached. Most land birds have six pairs of ribs, but a hummingbird's rib cage has eight pairs, affording the vital organs a greater degree of protection. The coracoid bone, which links the vertebral column with the apex of the sternum, is unusually strong in hummingbirds. It is attached to the sternum by a shallow ball-and-socket joint—an arrangement unique to hummingbirds and their close relatives, the swifts. Like swifts, hummingbirds do not walk, and the very short leg bones (femur, fibula, tibiotarsus, and tarsometatarsus) reflect this. However, four relatively long toes—three facing forward, and a fourth (the hallux, or hind toe) facing back—enable hummingbirds to perch easily on twigs and small branches.

Bones of the wing

Like other birds, a hummingbird has six main bones in each wing. Those nearest the scapula, or shoulder—the humerus, radius, and ulna—are relatively short compared with those of most other birds, while those farthest away (the second digits) are relatively long. This structure helps a hummingbird perform its amazing feats of hovering and flying backward.

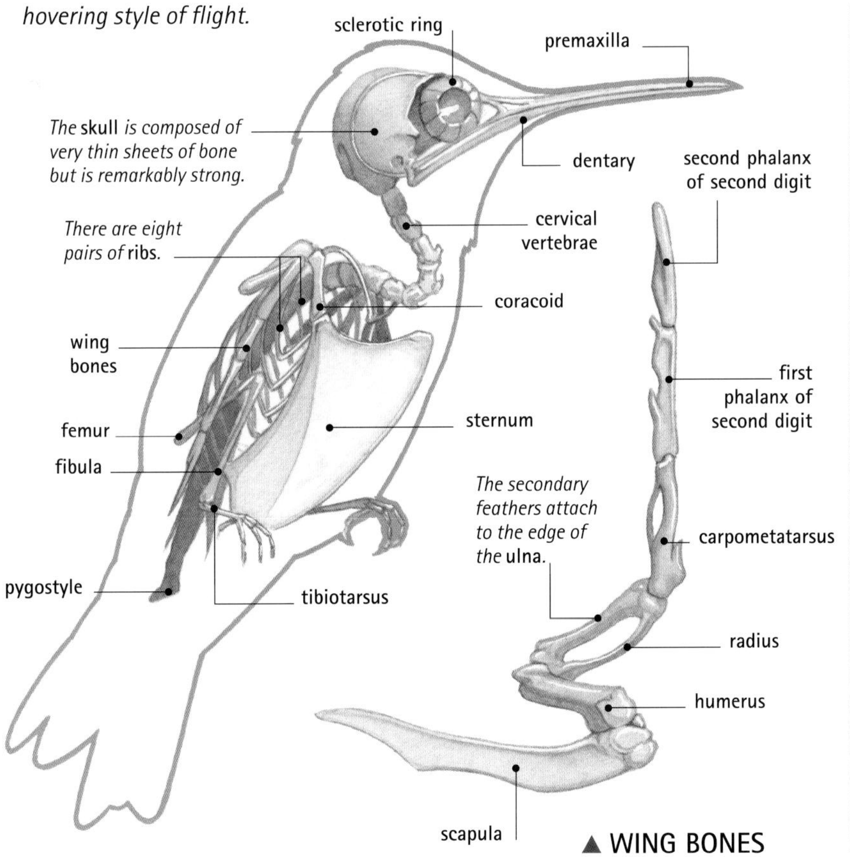

▼ **Ruby-throated hummingbird**
In common with other hummingbirds, this species has a deep keel on the sternum to attach flight muscles. The short "arm" bones and long "finger" bones help produce its hovering style of flight.

▲ WING BONES

EVOLUTION

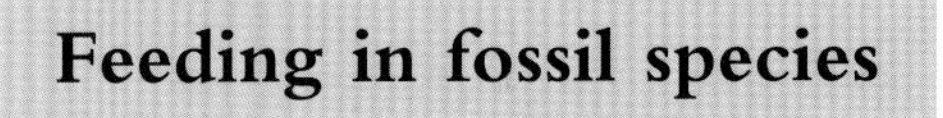

Feeding in fossil species

The fossilized skeletons of hummingbirds found in Central America, Europe, and Asia reveal that they once lived over a far greater range than they do now. The fossils also give scientists clues to how hummingbirds have evolved over millions of years.

Scientists once believed that these ancient hummingbirds were exclusively insectivorous birds that probed vegetation for tiny insects. On this theory, hummingbirds evolved their nectar-feeding diet over millions of years. However, two of the oldest known hummingbird fossils were found in Germany, in rocks 30 million years old. The fossils, of the species *Eurotrochilus inexpectatus*, have a long bill and wing bones similar to those of modern hummingbirds, suggesting that they were nectar feeders and capable of hovering. So hummingbirds similar to those alive today evolved much longer ago than was once thought.

Muscular system

As in other animals, all movement in hummingbirds is powered by muscles that are connected to parts of the bird's skeleton. The general pattern of muscles within a hummingbird's body is the same as in other birds, but the flight muscles are larger, relative to its body weight. The bones of the spine do not require extensive back muscles for support, and most of the muscle mass is concentrated near and below the bird's center of gravity, sending out long tendons to control the movement of the outer wings, legs, and feet.

The two pectoralis major muscles connect the sternum and keel with the humerus and clavicle bones and are responsible for the wings' downstroke. The two supracoracoid muscles connect the sternum with the dorsal surface of the humerus and are responsible for the upstroke of the wings, which—uniquely in hummingbirds—is as powerful as the downstroke. Together, these two pairs of muscles make up at least 30 percent of the hummingbird's overall weight, more than in any other bird.

CLOSE-UP

Muscle cells

Muscle fibers are composed of long, thin cylindrical cells full of the mechanisms required to convert chemical energy into movement. They are large compared with most other cells. The fibers are arranged parallel to each other and usually lengthwise along the muscle. A sheath of connective tissue (collagen) surrounds individual muscle fibers. Blood vessels, motor neurons (nerves that cause muscle fibers to react), and other nerves wind between the fiber bundles.

Unusually, the four muscles responsible for the flight power strokes contain all red muscle fibers and no white. (In other birds, these muscles contain at least some white fibers.) There are several kinds of muscle fibers. The red, or slow-twitch, fibers that make up hummingbird flight muscles are very resistant to fatigue. They contain more carbohydrate (glycogen) granules and fat (lipid) droplets than white (or fast-twitch) muscles, and there are more blood capillaries to supply oxygen to the muscle. The cells of red muscle fibers also contain more mitochondria—the parts of cells responsible for converting nutrients into energy—which contain the enzymes required for the oxidation of the carbohydrate and fat nutrients. A combination of more fat and carbohydrate fuel and more oxygen enables the red muscles to run and run. If hummingbirds' mitochondria occurred at the same density as those in the cells of mammalian muscles, the mitochondria would make up 70 percent of the flight muscle. In fact, the mitochondria in the flight muscles take up only half this space because they are packed together more tightly than in mammals' muscles.

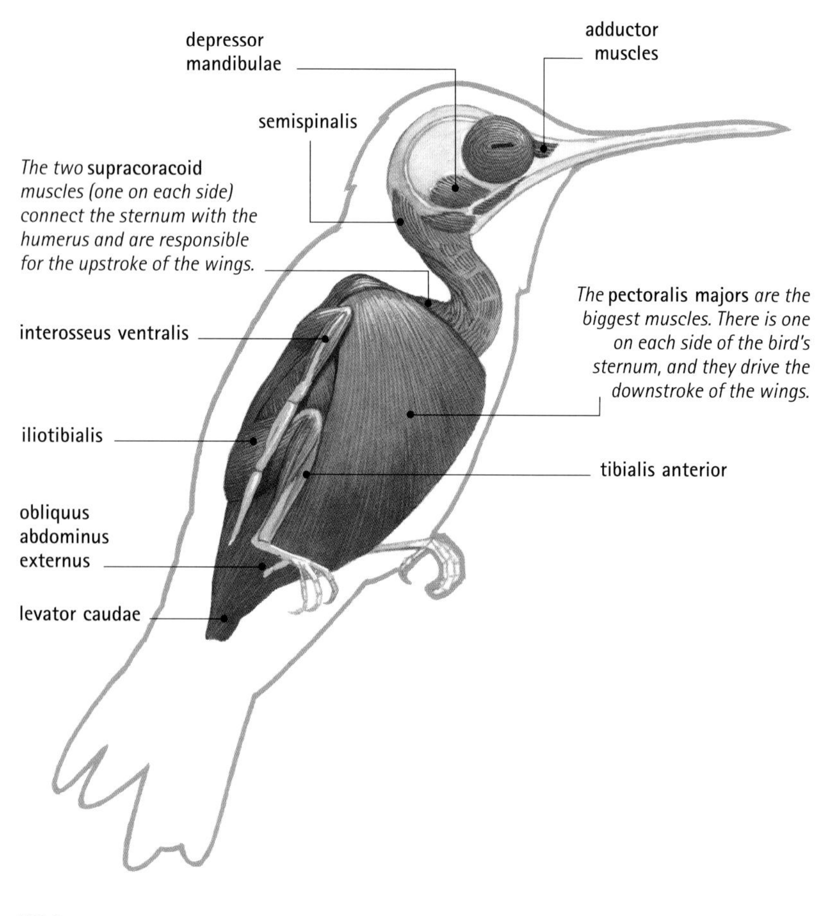

▼ **Ruby-throated hummingbird**
Two large pairs of muscles—the pectoralis major and the supracoracoid—provide the power for the downstroke and upstroke of the wings in flight.

Nervous system

In vertebrates, the brain and spinal cord form the central nervous system (CNS). The brain sends messages to the body and receives information through the peripheral nervous system (PNS). This joins the CNS at the spinal cord, which is protected from damage by the bones and membranes of the backbone.

The hummingbird's PNS consists of 38 pairs of spinal nerves, which run from all parts of the body to the spinal cord. Twelve pairs of spinal nerves join the spinal cord in the neck (cervix); 8 pairs join in the upper back (thorax) region; 12 pairs join the spinal cord of the lower back; and 6 pairs serve the tail, or caudal, region. The last two cervical spinal nerves and the first three thoracic spinal nerves form the brachial plexus, from which nerves run into the wings. Two sciatic nerves also run from the spinal cord down into the hummingbird's thighs. Many small nerves from each of the major ones ensure that every part of the bird's body can receive signals from and send signals to the brain.

A hummingbird's brain is among the largest, relative to size, of any bird's, weighing up to 4.2 percent of total body weight. The part of the brain that controls learning (the cerebrum), is relatively small compared with this part in birds such as parrots and jays, but the area that controls muscle activity and instinctive behavior (the cerebellum), is bigger than average for a bird.

A hummingbird's most important sense is vision. Its eyes are about 0.3 inch (0.8 cm) in diameter and are on the sides of the head, allowing the bird to see straight ahead and to the side. The eyes have relatively large optic nerves, and the retina has a very high density of rods (light sensors) and cones (color sensors).

Taste and hearing

The sense of taste in hummingbirds is sufficient to allow discrimination between solutions with high and low concentrations of sugar. Without this ability, a hummingbird could waste valuable time feeding on sugar-free liquid. A hummingbird probably has no sense of smell, since there are no olfactory nerve endings in the nasal cavities, but it probably has an acute sense of hearing. Hummingbirds will respond to the quiet calls of another hummingbird or the sound of its wings beating.

◀ *Hummingbirds have very sensitive color vision, which helps them identify the bright colors of nectar-bearing flowers.*

Circulatory and respiratory systems

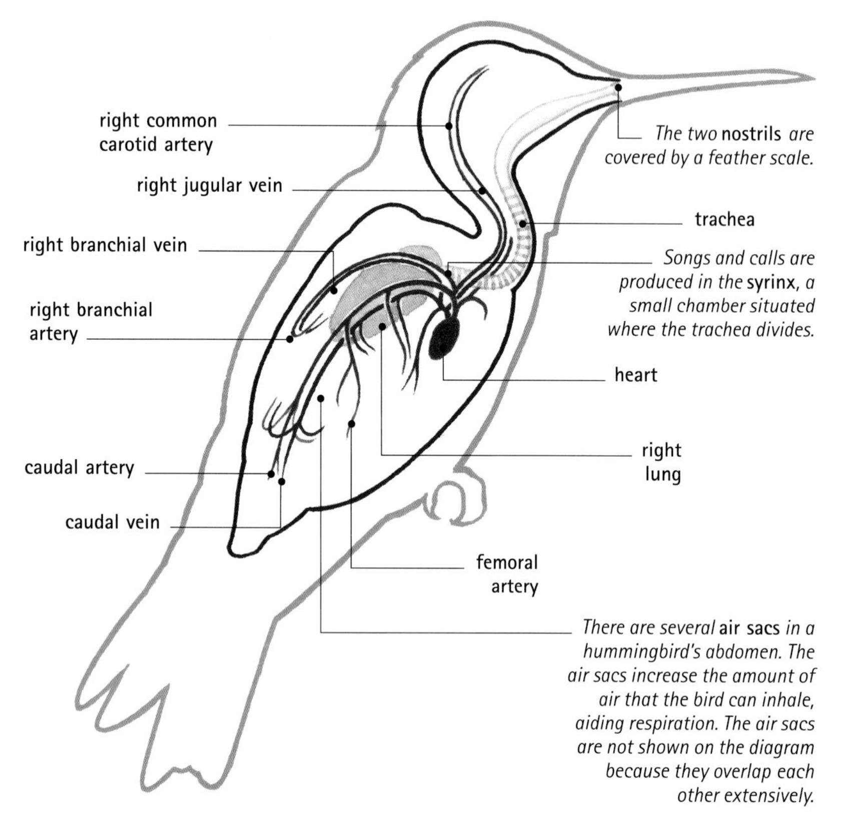

▲ Ruby-throated hummingbird
To provide the oxygen needed for flight, there is a high concentration of red blood cells, and air sacs supplement the lungs' capacity.

A resting ruby-throated hummingbird breathes around 250 times a minute. Air enters the respiratory system through the nostrils at the base of the bill, then enters the small, paired lungs. The lungs are connected to a system of air sacs that lie in spaces not occupied by other organs. Birds' air sacs increase the area over which gas exchange can occur, and thus make respiration more efficient than in mammals, which lack air sacs. Gas exchange involves the passage of oxygen from air in the lungs and air sacs into the blood for transport to the bird's cells, and the movement in the other direction of waste carbon dioxide. When a hummingbird flies, pressure from its muscles forces air in and out of the air sacs, thus maximizing airflow when the bird is most active and most in need of oxygen.

The circulation of oxygen around a hummingbird's body is extremely efficient. There are two reasons for this: oxygen is carried by erythrocytes, or red blood cells, in the bloodstream, and hummingbirds probably have the highest concentration of these cells of any animal; also, hummingbirds have the largest heart relative to body weight of any endothermic organism. The heart of a ruby-throated hummingbird accounts for around 2.5 percent of the bird's total weight, and the heart beats up to 1,260 times per minute—only some shrews have a faster rate. Even at rest, its heart rate is 500 beats a minute. As with other birds and mammals, the heart has four chambers. The left side receives oxygen-rich blood from the lungs and pumps it around the body; the right side receives deoxygenated blood from the body and pumps it back to the lungs.

The complex network of arteries and veins in a hummingbird's body carries nutrients and wastes (as well as oxygen and carbon dioxide) to and from organs, glands, and muscles all around the body. For example, nutrients and minerals are carried directly from the walls of the intestines to the liver via the hepatic portal vein. Inside the liver, the portal vein divides into a series of tiny blood vessels called capillaries to supply the liver tissues. Excess carbohydrates and fats are stored within the liver before being distributed throughout the body. They are carried out of the liver in another set of capillaries into the vein that transports blood away from the liver, the hepatic vein. The kidneys remove the waste products of metabolism from the blood.

CLOSE-UP

Torpor

In an environment cooler than their body temperature, small birds such as hummingbirds lose heat rapidly. Hummingbirds also lack down feathers for insulation, so they can maintain body temperature only by increasing heat production. To save energy during especially cool nights, hummingbirds may allow their metabolic rate to fall dramatically—to as little as to one-fiftieth of its normal rate—as they enter a state of torpor. Once a hummingbird is torpid, its heartbeat may be just 50 beats a minute, its body temperature may be 36°F (20°C) below normal, and the rate of water loss by evaporation may be just 10 percent of normal. Breathing is irregular, and there are long periods of nonbreathing.

Digestive and excretory systems

Hummingbirds eat some insects, but nectar from flowers makes up the overwhelming bulk of their diet and provides them with the energy they need to remain active. Studies of 15 North American hummingbird species showed that foraging from flowers takes up around 86 percent of their feeding time. The remainder of the time they hunt insects, which provide their protein requirements.

Hummingbirds insert their bill into the corolla of a flower and extend the tongue to reach the nectar. They use their tongue to lap up the sugar-rich liquid; they do not use their tongue and bill as a straw to suck out the nectar. Hummingbirds with a long bill—for example, sword-billed hummingbirds—can reach nectar in long, trumpet-shaped flowers, but those with a short bill have to feed from flowers with smaller corollas.

After feeding, nectar and other food passes along the bird's esophagus, a tube that runs down the neck and into the crop. The crop is an enlarged section of the esophagus that can stretch and store food for a short period. Sucrose stored in the hummingbird's crop provides a backup energy supply to keep it alive during the night when it is not able to feed. From the crop, food passes into another tube, called the proventriculus, where chemical digestion begins. The next stage in the passage

CLOSE-UP

A long tongue for feeding on nectar

When a hummingbird is not feeding, the posterior (rear) part of the tongue is curled around the eye sockets in a structure called the hyoid apparatus. This apparatus, which is partly bone, can extend the tongue well beyond the tip of the bill to reach deep into tubular flowers. The anterior (front) half of the tongue is divided into two, and each part is fringed, making it more efficient at lapping up nectar. A ruby-throated hummingbird extends and contracts its tongue up to 13 times a second when feeding, to maximize its nectar consumption.

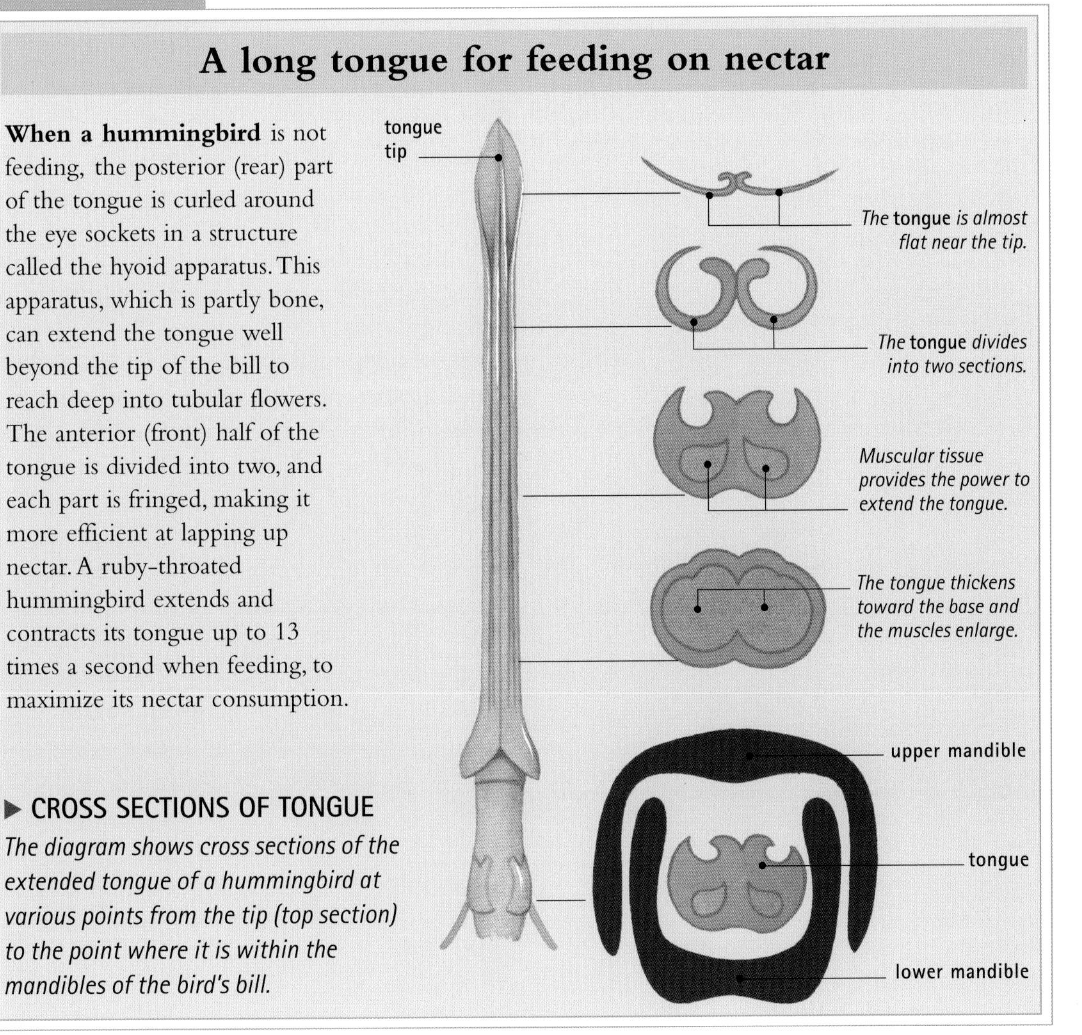

▶ CROSS SECTIONS OF TONGUE

The diagram shows cross sections of the extended tongue of a hummingbird at various points from the tip (top section) to the point where it is within the mandibles of the bird's bill.

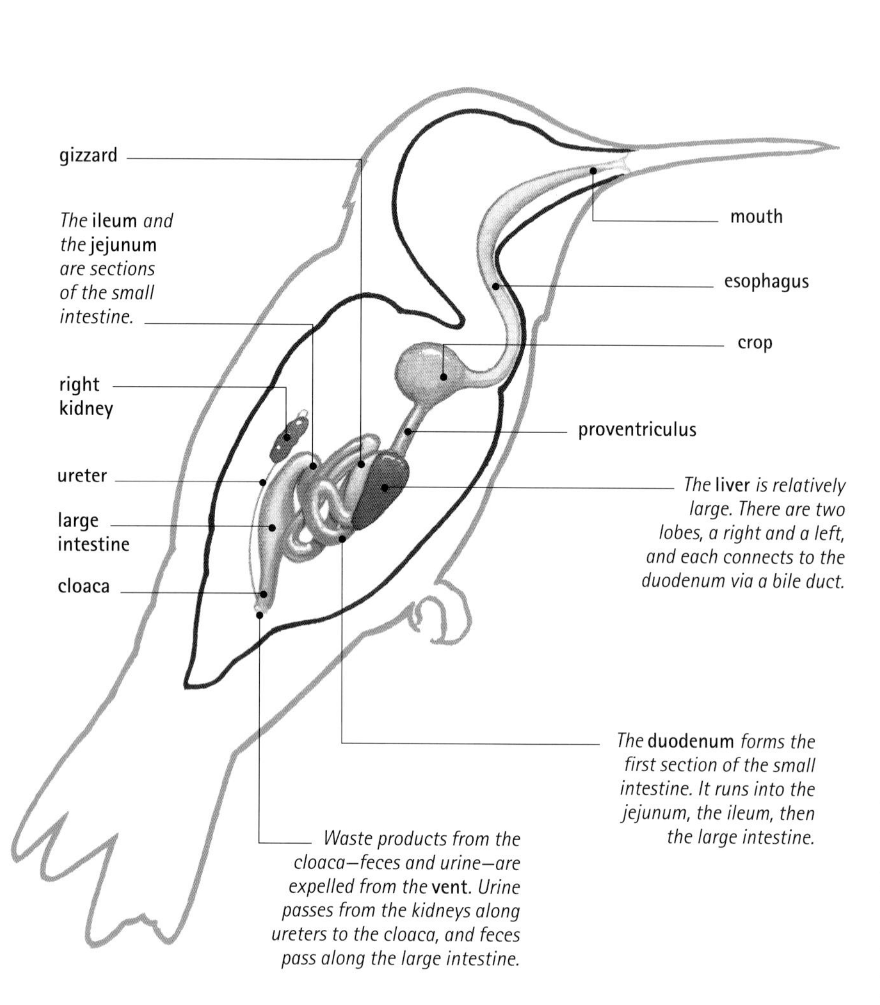

▲ Ruby-throated hummingbird
The main elements of a hummingbird's digestive and excretory systems.

of food is into the gizzard, a tough, muscular organ where the digestion of harder fragments of food—for example, the wings of insects—occurs. Bile from the liver and enzymes from the pancreas pass along the bile and pancreatic ducts respectively and then into the duodenum. The chemicals enable the digestive process to progress further: bile helps break down fats and carbohydrates, while the pancreatic enzymes break down carbohydrates, fats, and proteins. Later, the nutrient products of digestion are absorbed through the walls of the small intestine into the circulatory system. The small intestine of an average-size hummingbird is about 2 inches (5 cm) long. Waste products from the digestive process are stored temporarily in the large intestine before being expelled from the cloaca.

Waste from the kidneys passes along the ureter to the cloaca, where it mixes with solid waste from the large intestine. As in other birds, the wastes from the kidneys and digestive system are expelled together. Hummingbirds have no bladder in which to store liquid waste; urine is expelled as it is produced.

Feeding and migration

A hummingbird must spend a large proportion of every day collecting food simply to stay alive. Each day an individual hummingbird will often consume more than half its total weight in food and may drink twice its weight in water. Studies on Anna's hummingbirds demonstrated that an individual male would need to collect nectar from more than 1,000 fuschia flowers daily to supply its energy needs.

Some hummingbirds migrate seasonally between higher and lower latitudes to take advantage of locations with better food supplies. A hummingbird's energy needs become even greater than usual if it migrates. It will need to store lipids (fat) in its body before making a long flight, for example, from northern California to Mexico. Much of this lipid supply is in the bird's liver; in the middle of summer, lipids make up only 15 percent of the weight of a ruby-throated hummingbird's liver, but just before migration this figure rises to 45 percent.

Most hummingbirds that migrate do so over land so they can find food easily during their long journey. Ruby-throated hummingbirds, however, migrate 500 miles (800 km) or more over the Gulf of Mexico every spring and return each fall. Crossing the open ocean, they have no opportunity to feed. Scientists have shown that 0.07 ounce (2 g) of fat provides all the energy they need to undertake a 26-hour flight at an average speed of 25 miles per hour (40 km/hr); that is enough to take a male ruby-throated hummingbird 646 miles (1,040 km) and a slightly heavier female 605 miles (975 km). This is far enough to make it across the Gulf of Mexico with a little to spare. If the hummingbird is not able to build up the necessary fat reserves before departing, or if it gets blown off course by a storm while crossing the ocean, it will perish before reaching the other side.

Reproductive system

Hummingbirds, like other birds, reproduce sexually. A hummingbird's reproductive organs are active only during the (usually short) reproductive season. During the nonbreeding months, the male's testes, and the left ovary of the female, shrink to a tiny fraction of their active size. If they remained large all year, they would be an added weight burden. The female's right ovary is tiny and does not function.

Before they copulate (mate), hummingbirds often participate in elaborate courtship displays. During copulation, the male perches on the female's lower back, and the cloacas of the two birds are brought together. Sperm passes from the male to the female, and fertilization of her eggs usually takes place in part of the oviduct called the infundibulum. Then begins a journey down the oviduct, which may take 24 hours or less. Over a period of about three hours, the egg receives a coating of albumen, or egg white, in the magnum region of the oviduct. The egg then moves to the isthmus region where it remains for around one hour; there, the shell membranes are deposited on the egg. Next, the egg receives its outer shell during a stay of about 20 hours in the uterus region. Finally, the egg passes through the vagina, into the cloaca, and into the outside world.

TIM HARRIS

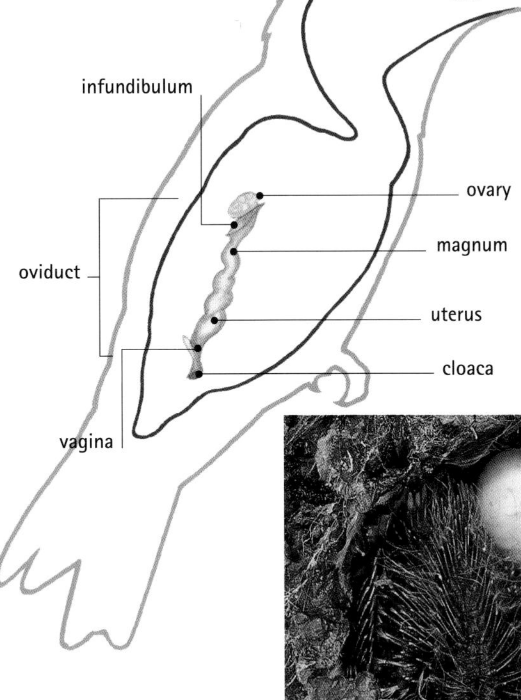

◀ **Ruby-throated hummingbird**
A female hummingbird has only one functioning ovary. The left ovary is shriveled.

▲ *Newly hatched hummingbird chicks are about the size of a honeybee and completely helpless.*

IN FOCUS

Incubation and eggs

Female hummingbirds usually lay two white eggs. In many species, the female begins to incubate immediately after laying the first egg, and the eggs hatch in approximately the same time sequence in which they were laid. The eggs are typically laid in the morning and at intervals of around 48 hours. The male plays no part in incubating the eggs, and the female cannot incubate all the time, since she has to spend part of the day searching for food. The eggs are incubated for 15 to 23 days, depending on the species and conditions, during which time the embryonic hummingbird develops inside the egg.

When they hatch, hummingbird chicks are naked and blind and are totally helpless. Despite this state, the newly hatched chick's crop is well developed, so it can ingest food almost immediately. The nestlings are incapable of feeding themselves, so the female pushes large amounts of food into them. The parent bird inserts her bill into the nestling's mouth and regurgitates food from her own crop into that of the young. Food consists of large quantities of tiny insects and smaller amounts of nectar. After a feeding, the chick's crop grows so large it pushes out the sides of the neck.

Most hummingbirds have a good covering of feathers by the time they are 12 days old, and they can fly by the age of 22 to 38 days. However, the mother may continued to feed her young until they are up to 65 days old.

FURTHER READING AND RESEARCH

Proctor, N. S., and Patrick J. Lynch. 1993. *Manual of Ornithology.* Yale University Press: New Haven, CT.

How Animals Work: www.sci.sdsu.edu/multimedia/birdlungs/

Operation Rubythroat: www.rubythroat.org

Hyena

ORDER: Carnivora FAMILY: Hyaenidae

Hyenas have an undeserved reputation as mean-tempered scavengers. In fact, they are highly specialized and skillful predators with extraordinary social lives. The digestive system is highly efficient: hyenas will crunch up the bones of much larger animals, and one species feeds almost exclusively on toxic termites.

Anatomy and taxonomy

The hyenas belong to a large group of meat-eating mammals called the carnivores (order Carnivora). Other carnivores include dogs, cats, bears, and ermines. Together, the four species of hyenas form the family Hyaenidae.

- **Animals** Animals are multicellular organisms that feed on organic matter from other organisms. They are able to move about and respond to stimuli.

- **Chordates** At some point in the life cycle of all chordates, the body is supported by a stiff rod called the notochord running along the back.

- **Vertebrates** The majority of chordates are vertebrates—animals with a supportive spine or backbone made up of articulated units called vertebrae. Vertebrates have a distinct head at the front and muscles in a paired arrangement on either side of the body.

- **Mammals** All mammals are homeothermic: that is, they maintain a warm, more or less steady body temperature. The body is usually covered with hair or fur. Females produce milk from mammary glands to nourish their young. Mammalian red blood cells have a nucleus, unlike those of other vertebrate groups.

- **Carnivores** Species of the order Carnivora eat meat, though many also eat plant matter. They have long, stabbing canine teeth and cheek teeth called carnassials, which have sharp edges adapted for shearing flesh and sinew. The group includes cats, dogs, and bears and smaller predators such as stoats and weasels.

- **Catlike carnivores** This group includes cats, civets, and hyenas. Most species are tropical or subtropical, and many are patterned with spots, stripes, or rosettes.

- **Hyenas** This is the smallest family of carnivores, with just four species. Modern hyenas are all superficially rather doglike, with a back that slopes downward from the

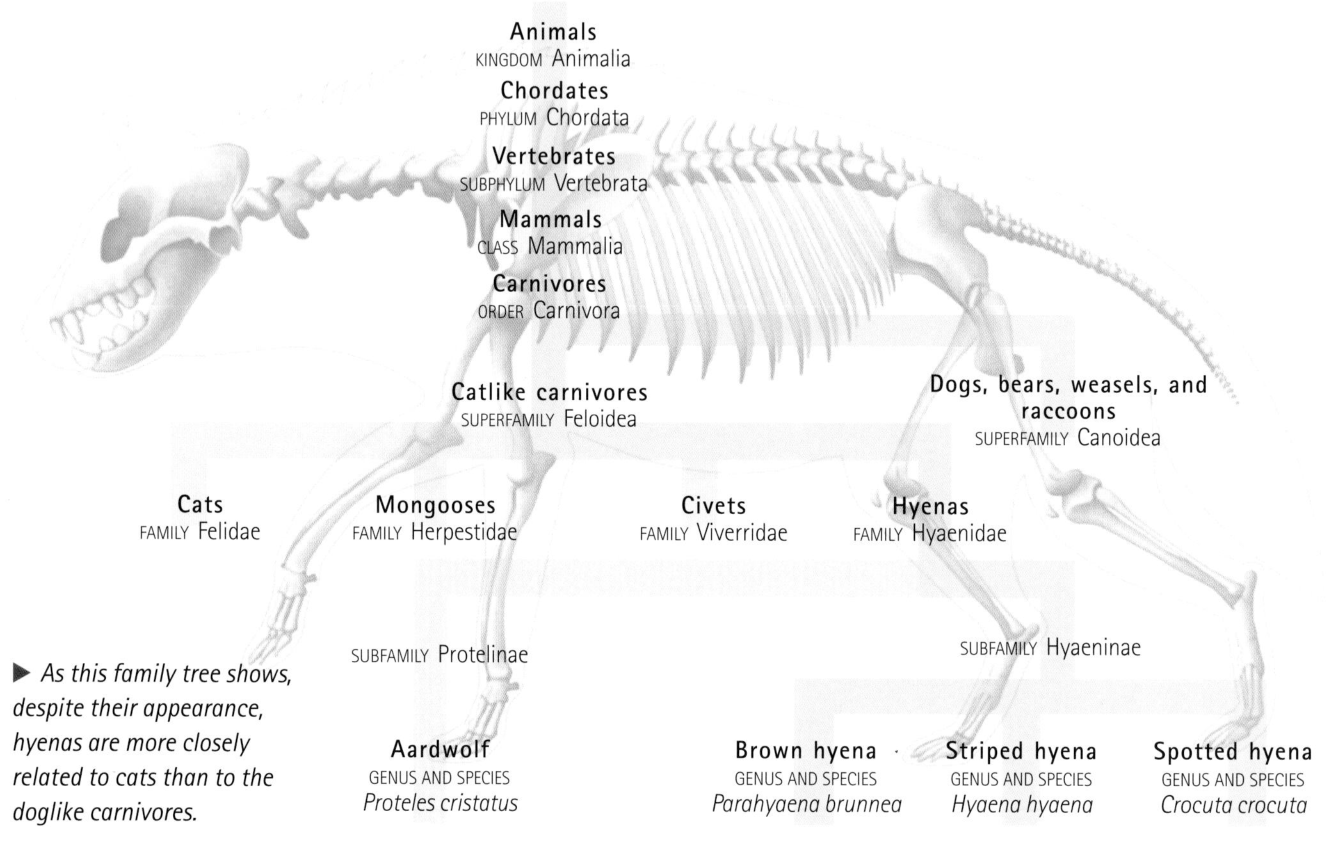

▶ *As this family tree shows, despite their appearance, hyenas are more closely related to cats than to the doglike carnivores.*

▲ *Spotted hyenas are the largest species in the hyena family. Like all hyenas, they have a doglike appearance but are actually more closely related to cats.*

shoulders to the rump. However, extinct members of the group were much more variable in form. The features that unite all members of the family, both living and extinct, are the detailed structure of the small bones of the inner ear and the possession of an eversible scent pouch under the tail. Brown, striped, and spotted hyenas are sometimes called true hyenas.

• **Aardwolf** A slender, long-legged hyena, the aardwolf lacks the massive head and jaws of its relatives. It has a yellowish gray coat marked with dark spots and streaks, and a crest of longer hair along the back from the nape of the neck. Aardwolves specialize in eating termites, and their cheek teeth, or carnassials, are peglike. Aardwolves live only in Africa.

• **Brown hyena** This species has pointed ears, and its brown coat is plain except for pale stripes on the legs and feet and a pale shaggy mane around the back and shoulders. Brown hyenas have a varied diet that includes carrion, live prey, insects, and fruit. They live only in Africa.

• **Spotted hyena** The largest of the hyena species, the spotted hyena is stoutly built with a shortish coat; a short, scruffy mane; and rounded, furry ears. It lives exclusively in Africa in socially complex clans and obtains most of its food by hunting. The bone-crushing premolar teeth are accompanied by bladelike carnassials for slicing meat.

• **Striped hyena** More widespread than its relatives, the striped hyena lives in northern Africa and through the Middle East to India. It has a shaggy coat, a longish mane, and pointed ears. Of all the hyena species, the striped is the most committed scavenger; all its cheek teeth, including the carnassials, are modified for crunching bone.

FEATURED SYSTEMS

EXTERNAL ANATOMY Hyenas are long-legged, long-necked, doglike animals with erect ears; a coarse coat; a shaggy crest, or mane; a bushy tail; and a front-heavy, slouching posture. *See pages 594–595.*

SKELETAL SYSTEM Hyenas have a large, heavy skull. The jaws are massive, with highly distinctive bone-crushing carnassial teeth. *See pages 596–597.*

MUSCULAR SYSTEM Hyenas are front-heavy, with a highly muscular head, neck, and shoulders. *See pages 598–599.*

NERVOUS SYSTEM Hyenas are intelligent, alert, responsive, and opportunistic. They possess excellent vision and hearing, and scent is very important in hunting and social behavior. *See pages 600–601.*

CIRCULATORY AND RESPIRATORY SYSTEMS Hyenas are endurance athletes, with a large heart and lungs and efficient circulation. *See page 602.*

DIGESTIVE AND EXCRETORY SYSTEMS Hyenas have an extremely strong stomach that, in most species, allows them to digest hard bones, horn, and teeth. *See page 603.*

REPRODUCTIVE SYSTEM The reproductive anatomy of spotted hyenas in unique in that the females' genitalia closely resemble those of males. *See pages 604–605.*

External anatomy

CONNECTIONS

COMPARE the strong front legs of a hyena with those of a *KANGAROO*.

COMPARE the bushy tail of a hyena with the almost hairless tail of a *RAT*.

Hyenas are front-heavy animals. The head and shoulders are large and powerful, and the forelegs are longer and more powerfully built than the hind limbs. The hindquarters are small and narrow and lower than the shoulders. One species of hyena, the aardwolf, has five toes on its front feet and four on the back. All the other hyenas have four toes on each foot. The toes are well padded, and each one bears a stout, nonretractile claw. The female of the spotted hyena is unusual for being larger than the male.

Hyenas look like dogs, but they are actually more closely related to cats. The doglike appearance has much to do with the fact that, like dogs, hyenas are cursorial mammals: their body is suited to a life of running. Hyenas' lifestyle has more in common with that of dogs than cats. The latter hunt by stealth, speed, and agility, while hyenas and most dogs (including wolves and jackals) specialize in strength and endurance. Hyenas can travel long distances with a tireless trotting gait or a steady lope.

A scruffy coat

Hyena fur is coarse and usually looks scruffy. All four species have a crest, or mane, of longer hairs along the back or around the shoulders, but this is less pronounced in the spotted hyena than in the other three species. The hair on the back of striped hyenas and aardwolves is three or four times longer than that elsewhere on

▶ **Striped hyena**
The striped hyena ranges in color from buff to straw and has distinctive blackish stripes. In the striped and brown hyenas, the males are slightly larger than the females. In the spotted hyena, however, the situation is reversed, with females being larger than males.

The thick **mane** *is made of coarse hair. When a hyena is threatened, the hairs become erect, giving the impression that the animal is much larger than it actually is.*

The **stripes** *run vertically on the body and horizontally on the legs.*

The **hindquarters** *of a hyena are noticeably weaker than the front quarters.*

The powerfully built **front quarters** *of a hyena give the animal a hunched appearance.*

Although hyenas have a doglike **face***, they are more closely related to cats.*

The striped hyena's **paws** *have sturdy nonretractile claws, which makes them suitable for a lifestyle of running on arid savanna and stony desert.*

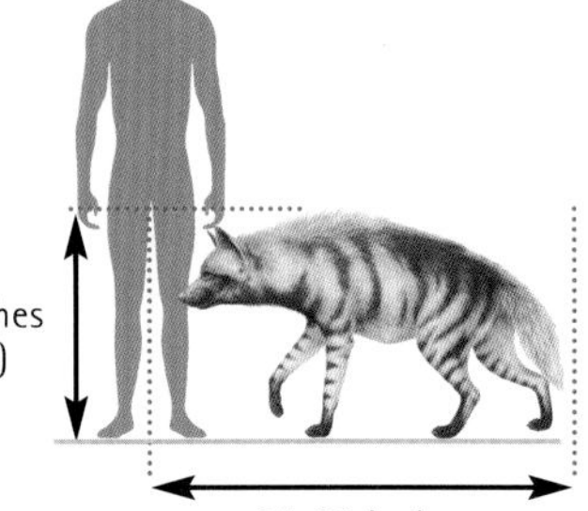

EVOLUTION

Hyena ancestors

The earliest ancestors of hyenas first appeared in Eurasia in the middle of the Miocene epoch about 22 million years ago. They were among the first catlike carnivores to abandon an arboreal (tree-living) lifestyle and adapt for life on the ground. They were omnivores (they ate meat and plant matter) and looked much like modern civets or mongooses. Over the next 7 million or 8 million years the ancestors of hyenas evolved into long-legged, doglike animals. During one period, hyenas were probably the dominant carnivores in Eurasia, with more than 30 different species. Most of these were generalist hunters, but they also included an aardwolf-like termite-eating specialist and the first hyenas with bone-cracking teeth. When true dogs arrived in Eurasia, and climate change drove many ancient carnivore lineages to extinction, the generalist doglike hyenas disappeared, leaving only the more specialized hyenas living today.

Ictitherium

The Ictitherium *is the earliest recognizable hyenid. Fossils of this animal have been found in rocks laid down around 15 million years ago.*

The tail *of the striped hyena is more bushy than that of the spotted hyena.*

the body and is made to stand on end when the animal is agitated. Combined with a threatening, back-arching posture, this erect fur can make the hyena look up to four times bigger than its actual size.

All hyenas have some kind of markings on their coat. Those of spotted and striped hyenas are dark brown to black on a buff to straw-colored background, or sometimes (in the spotted hyena) a reddish brown background. The muzzle is usually dark. The aardwolf has a variable pattern of stripes, streaks, and bars on its pale body and legs. Some scientists believe these markings mimic those of the much larger striped hyena, thus tricking potential predators into thinking the aardwolf is a stronger animal. However, a more likely explanation for the markings is that both species have retained the coat pattern of their common ancestor. Thus, although there is no fossil evidence, it is fair to assume that the common ancestor of all hyenas had a striped coat. In the spotted hyena the stripes have broken up into irregular spots, whereas the brown hyena's stripes have spread to cover most of the body. Some brown hyenas are uniformly chocolate brown, but most have a pale "cape" around the head and shoulders and light bars on the legs and feet. All four species have a long-haired, bushy tail, though the fur of a spotted hyena's tail is often rather sparse and straggly.

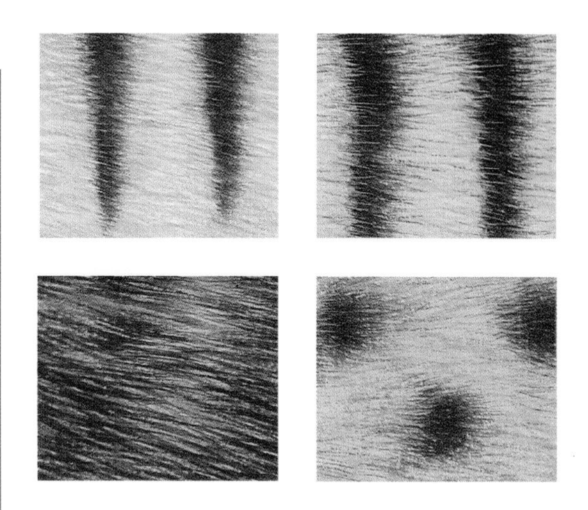

▲ **Fur markings**

Clockwise from top left: the aardwolf, striped hyena, spotted hyena, and brown hyena.

CLOSE-UP

Scent pouch

All four hyena species have large glands that secrete an oily scented paste into a pouch just under the tail. Scent is a very important part of social behavior. To mark its territory, a hyena turns its pouch inside out and straddles stems of grass, leaving behind neatly applied daubs of strong-smelling creamy-white paste. Native peoples in Africa refer to the secretions as "witches' butter." In striped and brown hyenas and the aardwolf, the pouch is also everted during social encounters, but in spotted hyenas subordinate individuals keep the pouch closed to avoid antagonizing the dominant members of their clan.

Skeletal system

CONNECTIONS

COMPARE the long neck and front legs of the hyena with those of a ***GIRAFFE***. Both have shoulders taller than the rump and a distinctive "rocking horse" gallop.

Hyenas have long legs suited to a lifestyle that involves a lot of running. Like most other carnivores, hyenas walk on their toes rather than the soles of their feet. This digitigrade stance allows an animal to run faster, since it effectively lengthens each leg and thus increases the length of the stride. The slender leg bones are the most vulnerable part of the otherwise very robust skeleton; therefore, rival hyenas often adopt a kneeling posture when fighting. The neck is also long, enabling the animal to reach the ground to feed and drink without crouching.

The dorsal vertebrae (bones of the spinal column) have long projections, or processes. These anchor the large shoulder and neck muscles, which in turn support the weight of the neck and head. The extra height at the shoulders emphasizes the slope of the back down to the hindquarters, even though the spine itself is more or less level.

Teeth

The teeth of striped, spotted, and brown hyenas are similar. There are three pairs of incisors at the front of each jaw and one pair of long canines. Hyenas' cheek teeth set them apart from all other mammals because the premolars are conical, with blunt tips for cracking bones. Like all carnivores, hyenas have carnassial cheek teeth with sharp edges for shearing tough meat, sinew, and hide. The teeth are sharpened by wear, and those in the preserved skull of an animal that died in its prime will cut paper as neatly as scissor blades.

The spotted hyena has the largest teeth, and this species has no trouble shearing the leathery hide, sinew, and bones of other animals. The carnassials of striped hyenas are better able to crush bones and hooves than to slice flesh; this species feeds more on scavenged carrion than do other hyenas. The dentary bone in the part of the jaw in which the cheek

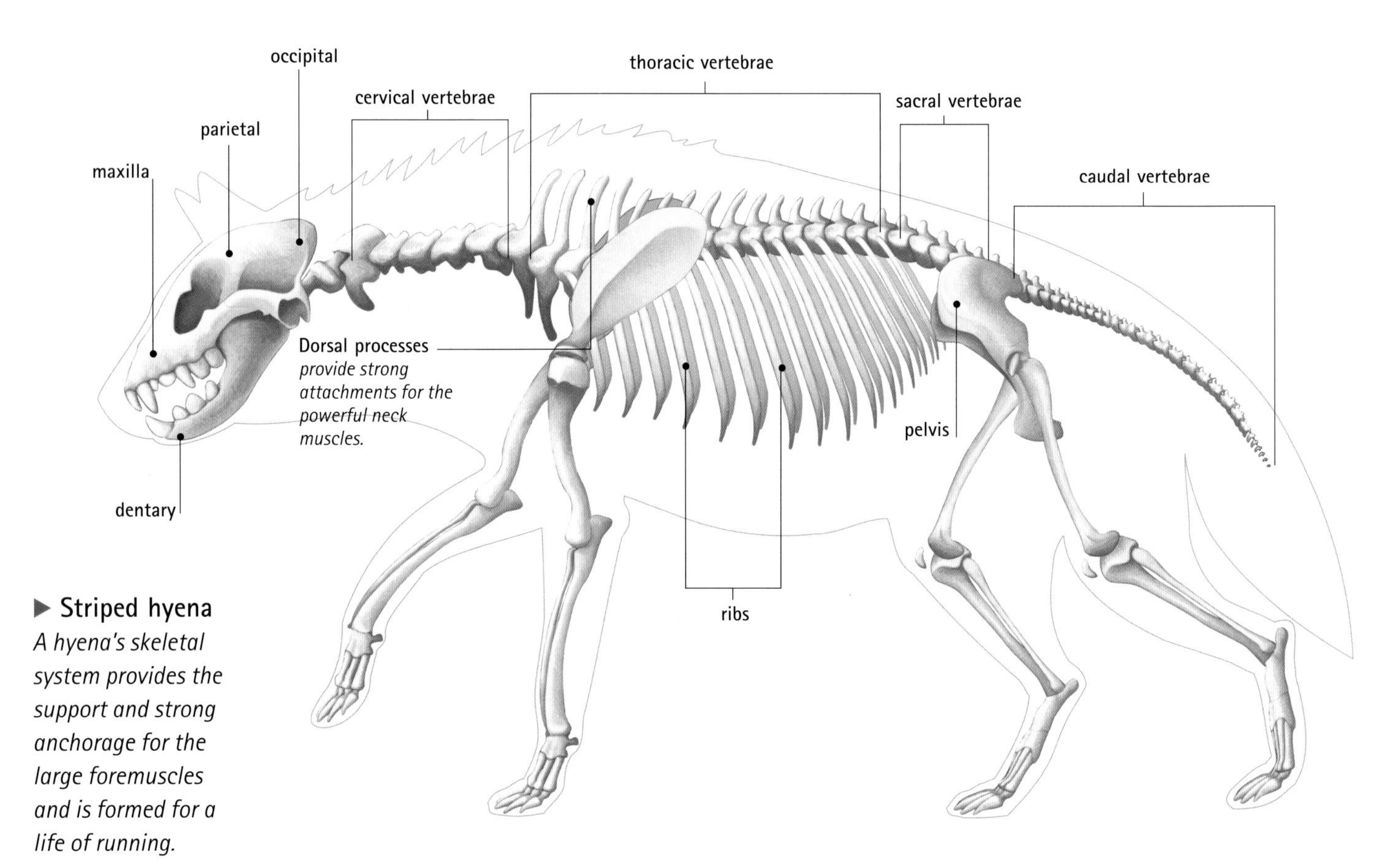

▶ Striped hyena
A hyena's skeletal system provides the support and strong anchorage for the large foremuscles and is formed for a life of running.

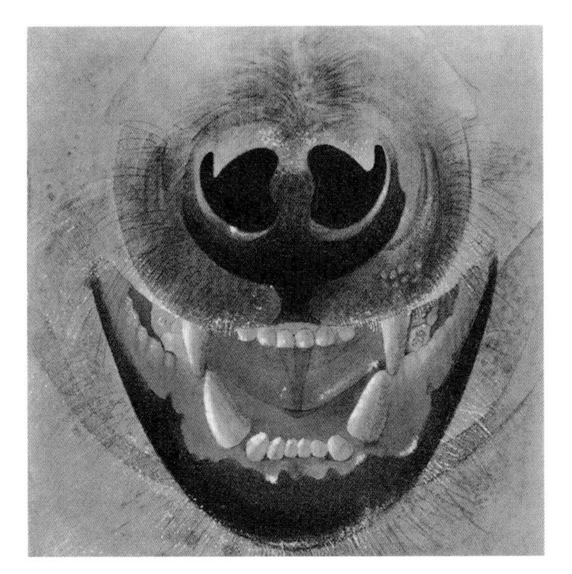

Muzzle
Hyenas have a dark brown to blackish muzzle with large canine teeth.

teeth are rooted is compacted and unusually hard. Normal bone would crack under the kinds of strain put on hyena jaws.

The aardwolf eats only termites, and its teeth are relatively unimportant in feeding; its carnassials and molars are hardly used. They are small, with wide gaps between, and they have no shearing edges or grinding surfaces. In contrast, the canines of an aardwolf are very well developed; aardwolves use them for fighting other aardwolves.

▲ *A spotted hyena displays the powerful slicing and crushing teeth it uses to tear into prey.*

Skulls

The spotted hyena's skull is large, short, and very robust. It has a very large sagittal crest (bony ridge along the top of the skull) and broad cheekbones. Both of these features are necessary to accommodate huge jaw muscles. The lower jaw itself is very large and heavy. The skulls of brown and striped hyenas are longer, with slightly less massive jaws. Spotted, striped, and brown hyenas have elongated sinuses (spaces) in the frontal and parietal bones of the skull. The skull of the aardwolf is much more dainty than the skull of the other hyenas, with longer, narrower jaws and a low sagittal crest.

◀ Spotted hyena skull
The spotted hyena has the largest and most robust skull and jaws. Canine teeth rip and kill prey, and the premolars crush bones.

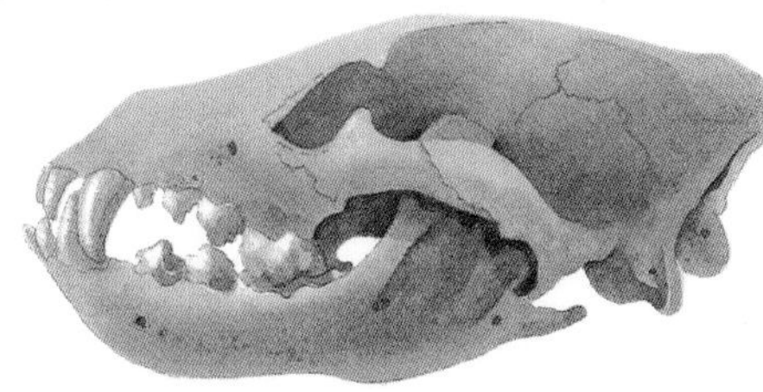

▶ Striped hyena skull
The striped hyena's skull is less robust, and its teeth are used more for cracking bones than slicing flesh.

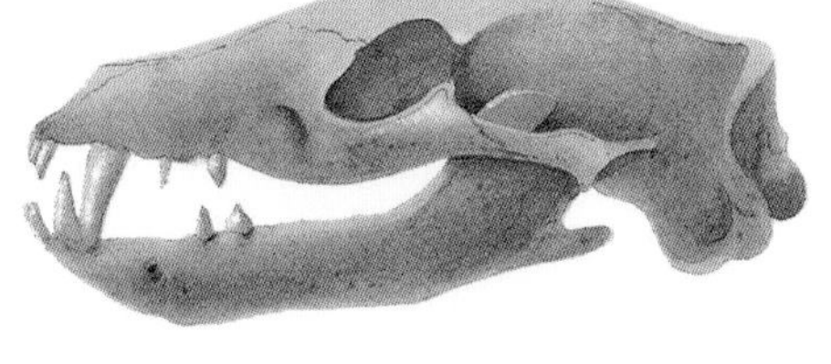

◀ Aardwolf skull
The long canine teeth are used in fighting, but the diet of termites does not require the fierce dentition found in other hyenas.

IN FOCUS

The hyena slouch

The downward slope of a hyena's back gives the animal a slouching posture, and the low carriage of the head gives the impression that it is habitually hunched or cringing. The hyena's reputation for cowardice has more to do with this posture than with its actual behavior. Hyenas are no less bold than other carnivores.

Muscular system

CONNECTIONS

COMPARE a hyena's muscular shoulders and neck with those of a *GRIZZLY BEAR*.

COMPARE the jaw muscles of a hyena with those of a *GIANT ANTEATER*. The anteater does not open its jaws to eat, and so its jaw muscles do not provide any crushing power.

The striking feature about the hyena physique is that all the bulk is at the front end. This is especially obvious in the spotted hyena. The head, neck, and shoulders are all powerfully muscled, and all play an essential part in the day-to-day business of chasing, killing, and consuming prey. Hyenas can also carry large carcasses in their jaws. Their ability to carry food is important in providing for young waiting back at the den, especially since hyenas do not regurgitate meat once it has been swallowed.

The head of true hyenas is very chunky and broad with a short muzzle. The huge muscles of the jaw are attached to the crest at the top of the skull. Most of the hyena's facial and neck musculature is concerned with operating the enormously powerful jaws. A hyena biting hard on a piece of bone can generate more than 11,400 pounds of pressure per square inch (800 kg per cm^2). That is equivalent to the weight of a small car pressing down on an area smaller than a person's little fingernail. The aardwolf's jaw muscles are less massive but can open the mouth very wide, exposing the long sharp canines used for fighting. The neck is very thick, heavy, and muscular in all species. The skin of the neck is thickened, protecting the underlying muscles from bites inflicted by struggling prey or other hyenas when fighting.

The rest of the body looks lightweight and weak compared with the head, neck, and shoulders. Hyenas sometimes have to travel 20 miles (32 km) or more in a night, and a heavy body would make this harder work. The disproportionate distribution of muscle in a hyena's body gives it the strength required to pull down prey as large as zebras and antelopes while maintaining a body light enough to run for many hours.

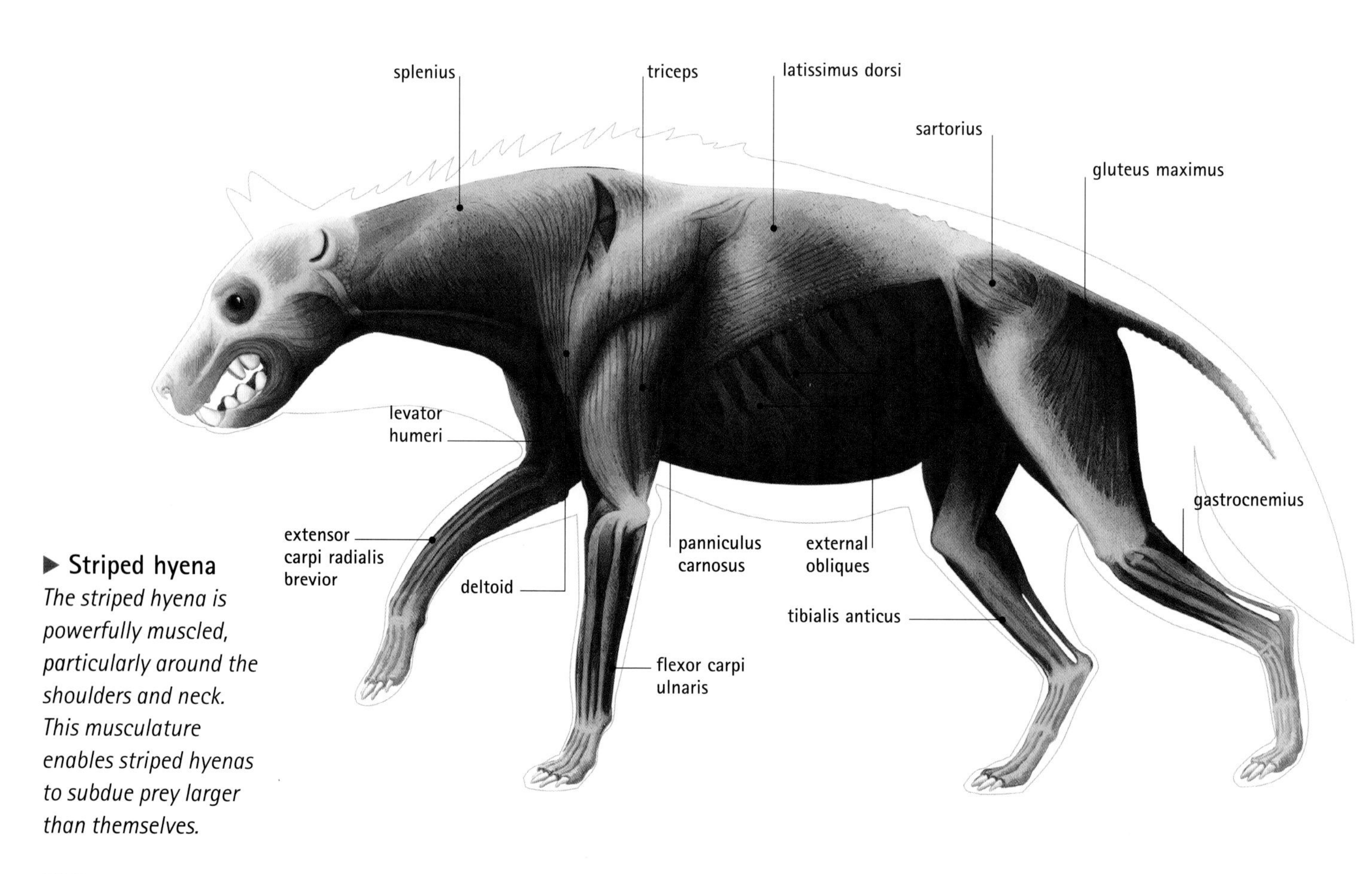

▶ Striped hyena
The striped hyena is powerfully muscled, particularly around the shoulders and neck. This musculature enables striped hyenas to subdue prey larger than themselves.

▲ *A hyena can run tirelessly for long distances in search of prey.*

The tail has its own musculature and can be held upright out of the way when the hyena is urinating, defecating, or leaving a scent mark, and in a variety of different positions to indicate mood. A hyena preparing to attack carries its tail very straight, whereas a submissive or frightened animal tucks it well between the legs. When excited, hyenas carry their tail curled forward over the back.

Digging a den

Digging is another activity that favors a front-heavy build, but because of their rather slender front legs, hyenas are not particularly efficient diggers. All four hyena species can dig their own burrows, or dens, but it takes them a long time. Whenever possible they will adapt existing burrows, especially those excavated by aardvarks.

Unlike many other termite-eating mammals, the aardwolf rarely digs for its food. Its claws, though stout, are no match for the concrete-hard fortifications of African termite mounds. An aardwolf will sometimes use its chisel-shaped lower incisor teeth to loosen hard packed earth and scrape it aside, but generally it prefers to feed on columns of harvester termites as they travel over the surface.

CLOSE-UP

The aardwolf's tongue

One of the largest muscles in a hyena's head is the tongue. In the true hyenas this is used for drinking, lapping up blood and other juices, rasping at flesh, and grooming the fur. In the aardwolf, the tongue is an enormous termite-harvesting device. The aardwolf's tongue is exceptionally long and broad, and during feeding it is covered in large amounts of runny saliva from very large salivary glands. The aardwolf's tongue is perfect for lapping up thousands of insects in a very short time.

Nervous system

CONNECTIONS

COMPARE a hyena's keen sense of smell with that of another animal that uses smell to track prey, such as a ***HAMMERHEAD SHARK***.

COMPARE a hyena's use of scent to communicate status and sexual condition with a ***HAWKMOTH***'s use of pheromones to find mates.

Hyenas are intelligent, adaptable animals. They are highly opportunistic and demonstrate keen judgment in the face of potential danger.

Hyenas have an uncanny knack of knowing when a prey animal such as an antelope or gazelle is sick or weak. To human eyes, all members of a relaxed herd of antelope may appear equally healthy, but when a pack of hyenas attack, they nearly always manage to single out the most vulnerable animal, whose frailty becomes obvious to humans only when it tries to flee.

Ever the opportunists, hyenas will hunt or scavenge at any time of the day or night, and even when they are sleeping part of the brain remains alert to sounds and scents that may signal a feeding opportunity. Their ears are large and mobile, especially those of the aardwolf. Hyenas have large eyes and are able to see well during the day and at night. In all species, however, the sense of smell is the most important. The part of the brain associated with smell—the olfactory bulbs at the front of the cerebellum—are well developed.

Hyenas use scent to communicate. Sniffing the anal pouch or genitals is an important part of the greeting ritual between individuals, and it may be that this is the way hyenas recognize different individuals. The personal scent of each animal is an identity badge that also provides a wealth of additional information about age, status, and sexual condition. When two hyenas meet, the mutual sniffing is equivalent to the typical conversation between two people who have just met: they introduce themselves, and inquire and provide answers about each other's health, status, and family. All this information is processed by the brain, and the hyena's subsequent behavior—dominant and aggressive or meek and submissive—will depend on what has been learned.

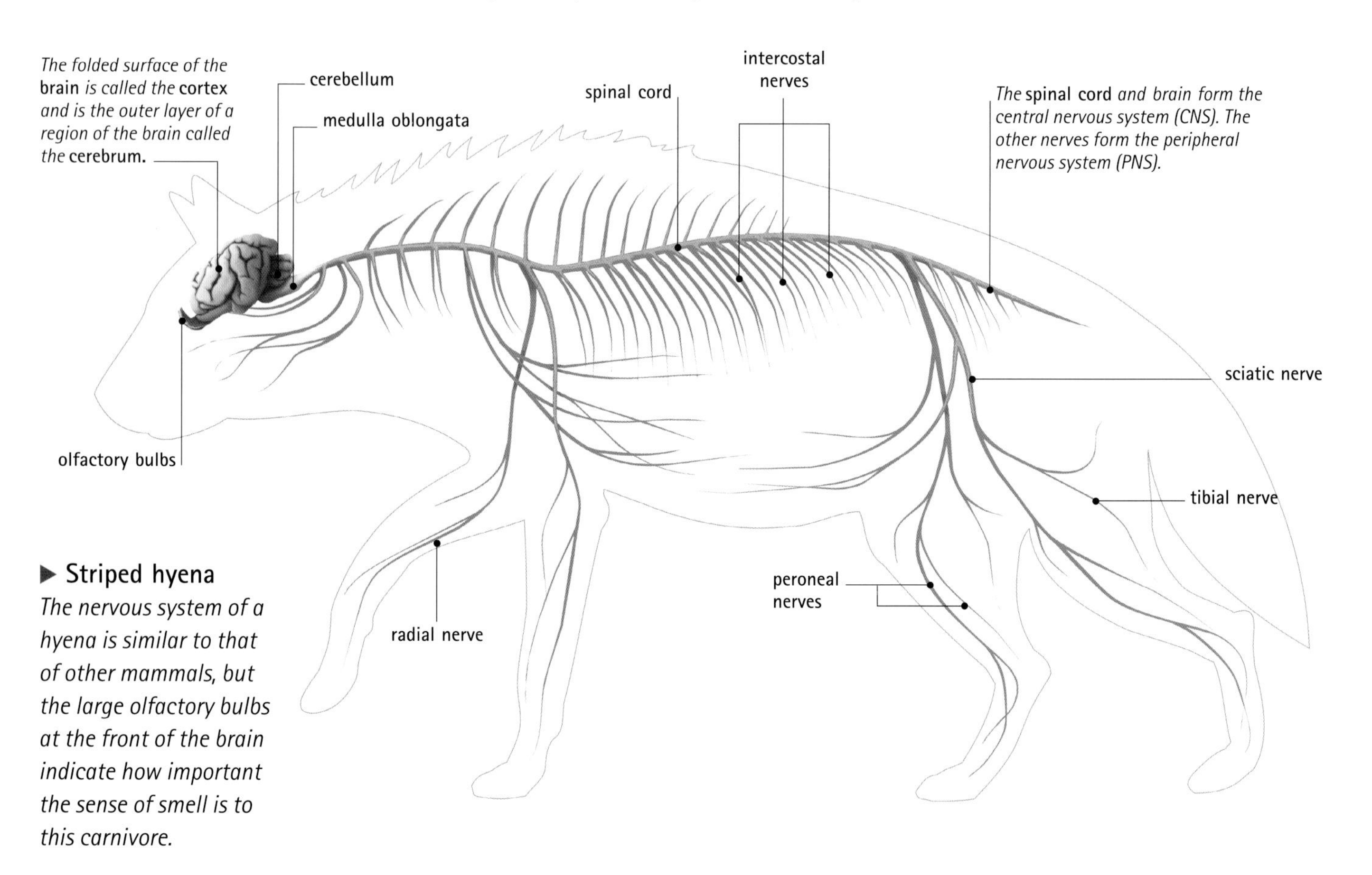

▶ Striped hyena
The nervous system of a hyena is similar to that of other mammals, but the large olfactory bulbs at the front of the brain indicate how important the sense of smell is to this carnivore.

▲ *When meeting, spotted hyenas smell each other's anal and genital region. This provides them with information about status and sexual condition.*

The scented paste from the anal gland that hyenas smear on grasses around their territory carries many kinds of socially important information. The scent changes and fades rapidly over time, so hyenas will be able to tell accurately how up-to-date the information is. Another hyena coming across the scent mark will investigate first by sniffing and sometimes tasting it. Unlike the sweet, musky scent produced by civets and used in the perfume industry, the scent of hyena paste is not pleasant to humans.

The brain, in particular the hypothalamus, is responsible for regulating the release of sex hormones, and in the spotted hyena in particular these have a profound effect, both physically and behaviorally. Spotted hyenas live in clans in which only the dominant females breed. Since hyenas are scavengers, these females must be extremely aggressive to protect their young and gain access to sufficient food. The hormone that promotes aggression is the steroid testosterone. This is the same chemical that promotes male reproductive development. Testosterone levels are often higher in dominant female hyenas than in males.

CLOSE-UP

Mystic powers

African folklore attributes great importance to hyenas. Their ability to arrive whenever lions make a kill or a large herbivore is sick or injured is believed by some to be due to supernatural powers. Parts of the hyena's body are used in medicines and charms, and some people believe these increase intelligence or impart some of the hyena's sharp senses.

Circulatory and respiratory systems

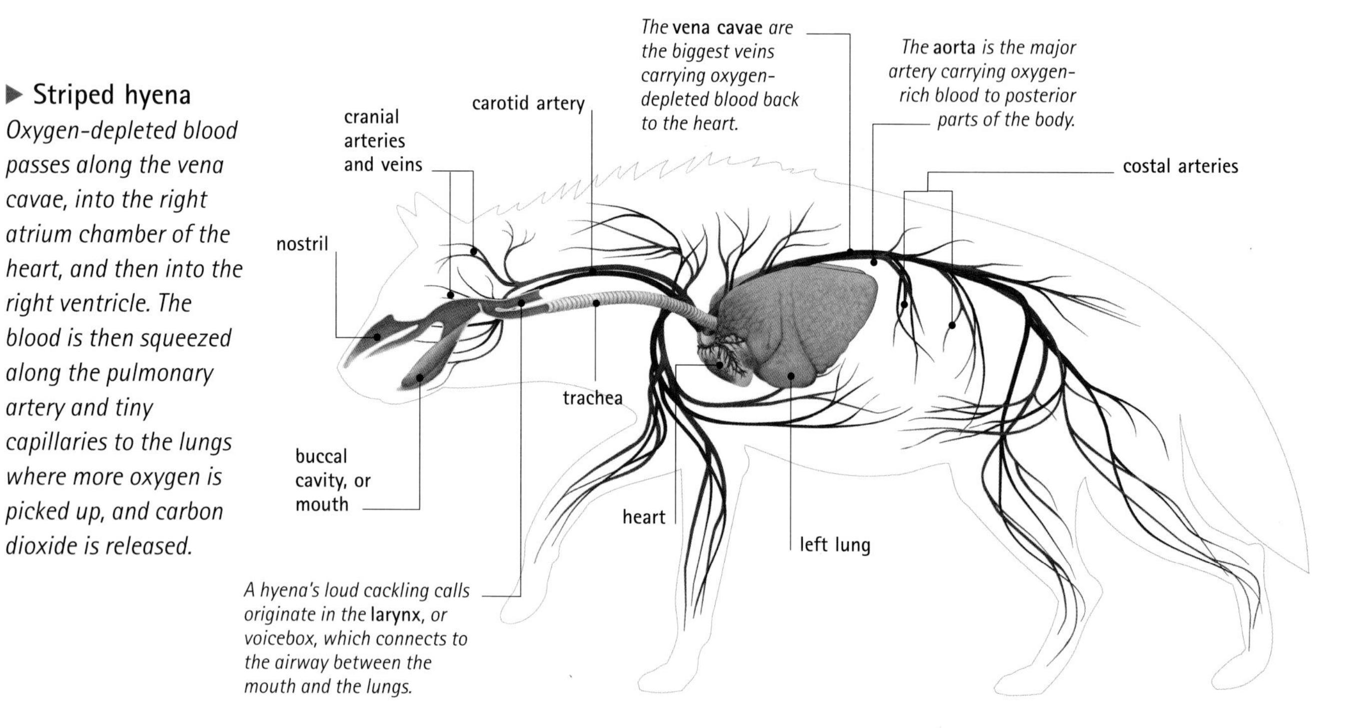

▶ **Striped hyena**
Oxygen-depleted blood passes along the vena cavae, into the right atrium chamber of the heart, and then into the right ventricle. The blood is then squeezed along the pulmonary artery and tiny capillaries to the lungs where more oxygen is picked up, and carbon dioxide is released.

Like all mammals, hyenas are endothermic (warm-blooded) animals. Blood is pumped from the slightly smaller right side of the heart to the lungs, where gas exchange takes place as the blood passes through many small capillaries. Oxygenated blood drains back toward the heart and is sent out around the head and body, this time via the larger left atrium and ventricle and a single aortic arch, on the left.

Hyenas can run fast for extended periods of time, especially when chasing prey. Pursuits may take place over 1 or 2 miles (1.6 to 3.2 km), and hyenas may reach speeds of up to 35 miles per hour (60 km/h). Hyenas' large lungs and efficient circulation enable them to sustain such efforts. In a flat run, a hyena breathes through its mouth. Having caught up with a large prey animal, the hyena must bring it to the ground and kill it; then the hyena's breathing switches to the nose, so the mouth is free for biting.

When a hyena has to chase and kill its own prey, most of its circulation is diverted to the muscles of the heart, lungs, and legs, rather than to the stomach and gut, so digestion and the absorption of nutrients does not begin until the animal has a chance to relax.

When a male hyena is sexually excited, blood is diverted to its penis, which has no supporting bone and thus relies on blood pressure to make it erect. The blood supply to the pseudopenis of female spotted hyenas is also increased during estrus, and the tissues around the urogenital opening relax greatly. The pseudopenis retracts and opens up, and the female is able to mate.

IN FOCUS

Laughing hyenas

The striped, spotted, and brown hyenas are also sometimes known as laughing hyenas because they make far-carrying, cackling calls when very excited. The sounds originate in the throat, where contractions of the larynx turn exhalations of air into what sounds like manic laughter. More common vocalizations are single yelps, whines, moans, and growls. Young hyenas give sharp bleating cries when begging for food. Hyenas, especially older male spotted hyenas, also produce a whooping call. A whooping male stands with his head lowered and his chest acting like a pair of bellows. High-ranking clan males are usually the loudest whoopers, and it seems that by drowning out the whoops of other males, they remind the dominant females of their high status, so that when the time comes to mate they will be favored.

Digestive and excretory systems

In the minutes after they make a kill hyenas eat incredibly fast to ensure that they get their share. Flesh and soft tissues are ripped from the prey and swallowed in large chunks. Competition over large kills is fierce, not only from other hyenas, but also from other species, in particular lions. It is a myth that hyenas always try to drive lions away from their kills. Much more often it is the other way around, and the agitated hyenas gathered around feeding lions are likely not to be the ones that made the kill in the first place.

The stomach juices of hyenas are much more acidic than those of other mammals. Meat and soft tissues are reduced to soup in no time, with no need for chewing or presoftening the flesh with saliva. Having been crunched up and swallowed, even large fragments of bone and teeth are completely dissolved within a matter of hours. The aardwolf does not swallow large chunks of bone, but it still needs a very strong stomach to break down the toxins in its termite prey into harmless chemicals. The ability to digest complex materials is something that evolved very early in the ancestors of hyenas and their relatives.

Hyena feces, in particular those of striped and spotted hyenas, dry into hard, chalky pellets; finely powdered calcium minerals are all that remain of the bones of carrion and prey once all the organic content has been absorbed. Hyenas are economical in their use of water but still need to drink regularly. In desert and arid conditions they supplement their fluid intake by eating succulent plants and watery fruits. The kidneys and bladder of hyenas are much like those of other carnivores, but the urinary duct of female spotted hyenas is unusual in that it extends the length of the pseudopenis that occurs uniquely in this species.

CLOSE-UP

Nothing wasted

Hyenas are the ultimate waste-disposal units. A pack of spotted hyenas can demolish an entire zebra or large antelope carcass in 15 minutes, leaving nothing but a the largest teeth and the bases of the horns. Hyenas also avoid the partially digested plant matter from a herbivore's large stomach, and they regurgitate balls of hair. Compared with hyenas, other carnivores are very wasteful. For example, lions usually eat only about two-thirds of the body of their prey, so it is not surprising that lion kills attract packs of hyenas eager to polish off the leftovers.

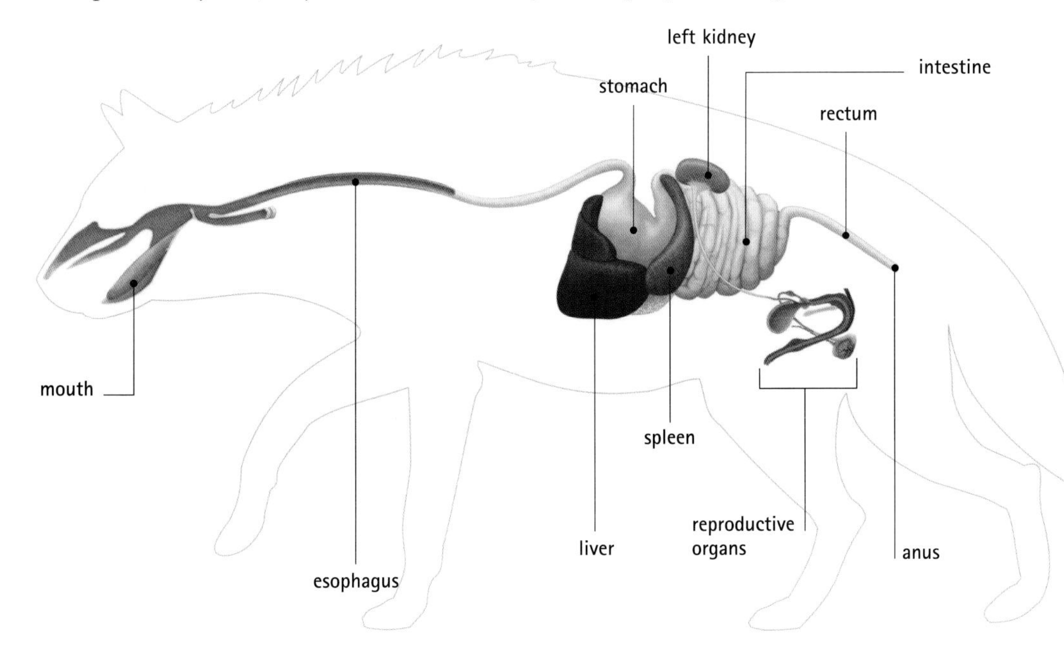

◀ **Striped hyena**
Important features of the hyena's digestive and excretory systems. In common with most other carnivores, a hyena's intestines are relatively short.

Reproductive system

CONNECTIONS

COMPARE the genitalia of a female spotted hyena with those of another mammal, such as a ***FRUIT BAT.***

COMPARE the barbed penis of a hyena with the similarly barbed penis of a ***LION.***

Striped and brown hyenas and aardwolfs have typical mammalian genitals. The genitals of female spotted hyenas, however, are highly unusual. Early scientific accounts proposed that the species was hermaphroditic, with each individual having both male and female reproductive organs. Biologists now know this is not true, but the misunderstanding is easy to understand: both male and female spotted hyenas appear to have a penis and scrotum. In females, both the "penis" (actually a greatly enlarged clitoris) and the pouchlike fatty swelling of the false scrotum are a result of the high levels of testosterone that begin circulating in the blood well before they are even born.

Although the external genitals of both sexes look like male reproductive organs, only males produce sperm. Immature females that have not yet had a chance to breed are thus very difficult to tell apart from males. However, once a female has given birth and suckled young, her teats remain enlarged, and her gender becomes more apparent. In males, the scrotum sheds its fur and darkens as the animal gets older. Female spotted hyenas have one pair of teats. Striped hyenas have two or three pairs; and brown hyenas and aardwolfs have two pairs.

CLOSE-UP

Genetics

Spotted hyenas live and hunt in clans of closely related individuals. Brown and striped hyenas hunt and scavenge independently, but their living arrangements are still communal. The relatedness of female clan members is important because even though not all will get a chance to breed, any cubs that are born will share at least some of the same genes. Thus by remaining in the clan and sometimes even helping the breeding female to feed her young (as do nonbreeding female striped and brown hyenas) the subordinate females maximize their own genetic stake in the next generation.

Female reproductive organs

Internally, the female reproductive organs of all four hyena species are similar. There are two ovaries, with oviducts leading into the twin "horns" of a more or less conventional carnivore uterus. The uterus and vagina are

▼ **MALE AND FEMALE REPRODUCTIVE ORGANS**
Spotted hyena
The external genital organs of the spotted hyena have been a source of some confusion to zoologists. The female's genitalia so closely resemble those of the male that the sexes are often difficult to tell apart.

penis
scrotum
anus

Adult male

The enlarged clitoris *forms a pseudopenis, which may be erected like a penis.*

The fused vulva *forms a false scrotum*

small nipples
anus

Subadult female approaching first estrus

The vaginal tract runs through the pseudopenis*. This makes mating and giving birth difficult: the male must insert his penis into the female's pseudopenis during mating, and the cubs must pass through this narrow organ when the female gives birth. The pseudopenis usually ruptures during birth and can take several weeks to heal.*

enlarged nipples
anus

Female that has borne young

▲ *A female spotted hyena feeding her cubs. Hyena cubs are born in an advanced state of development, with open eyes and many teeth.*

lined with smooth muscle. The muscle is best developed in the spotted hyena because the large pups have to be pushed out through the unusual birth canal, which curves toward the front of the animal.

Spiny penis

Unlike some other mammals, male hyenas do not have a penis bone, or bacculum. However, the tip of the penis, the glans, bears numerous backward-pointing spines, which help secure the penis inside the female during mating. Hyena pairs are locked together for some time. The barbed penis is a feature that male hyenas have in common with cats such as lions.

Female spotted hyenas receive less help to rear their young than brown or striped hyenas, but their litters are smaller: usually just one or two pups, compared with up to five in brown and striped hyenas. Heavily pregnant or nursing females cannot leave the area of the den to hunt, but such is the supremacy of breeding females within the clan that they are always allowed to eat their fill from a kill before the others take a turn, even if they were not involved in the hunt.

AMY-JANE BEER

IN FOCUS

Sibling rivalry

Having been flooded with testosterone during their embryonic development, spotted hyena pups fight almost from birth. Aggression is greater between siblings of the same sex, particularly between twin sisters. Since females tend to stay with the clan into which they are born, eliminating potential competition at an early stage increases a young female's chances of rising to dominance and thus getting a chance to breed herself when she is older. Around one-quarter of pups die of starvation as a direct result of bullying by their siblings.

FURTHER READING AND RESEARCH

Macdonald, David W. 2006. *The Encyclopedia of Mammals*. Facts On File: New York.

Nowak, Ronald M. 1999. *Walker's Mammals of the World*. Johns Hopkins University Press: Baltimore, MD.

Immune, defense, and lymphatic systems

From the instant an animal is born, and all through its life, its body is under threat from disease-causing organisms, or pathogens. Humans and other vertebrates come under attack; so do invertebrates. Pathogens are very diverse in size, in what they contain, and in what they have evolved to do. Since animals' bodies are not impenetrable, they are invaded almost continually. Yet most life-forms manage to stay healthy most of the time. This is because of their immune system, a complex system of physiological mechanisms that has evolved to identify, attack, and destroy most enemies. Most of the time, the immune system is successful.

The enemy

There are an enormous number of disease-causing organisms that can enter and infect the body of others. Among them are bacteria, single-celled organisms without a nucleus; protozoans, single-celled organisms with a nucleus; viruses, minute packets of DNA and protein, which do not have the other structures or equipment of a cell; various types of fungi; and larger organisms such as parasitic worms.

SYSTEM HIGHLIGHTS

INNATE IMMUNE SYSTEM Innate immunity is the system of immunity which animals are born with and which is not altered by their experience of fighting pathogens. This system includes the protective effects of the skin, phagocytic cells, and mucus on membranes. *See pages 609–611.*

ADAPTIVE IMMUNE SYSTEM The adaptive immune system evolves as a result of an animal's experience of pathogens. The two main groups of adaptive immune cells are B cells and T cells. The adaptive immune system produces antibodies that destroy invading pathogens. These antibodies evolve by a process of mutation, becoming increasingly effective at fighting off particular infections. *See pages 612–617.*

LYMPHATIC SYSTEM A system of ducts that permeates the body and recycles and cleans body fluids. This fluid contains immune cells that identify and destroy foreign antigens. *See pages 618–621.*

AUTOIMMUNITY AND GENETIC DISORDERS Sometimes the immune system malfunctions and attacks an organism's own cells. *See pages 622–623.*

Pathogenic organisms cause harm in a variety of ways. Some poison the body with the toxins they produce: for example, the bacteria that cause diphtheria in humans live in the trachea and produce a dangerous toxin that locally disorganizes the function of the tissues. Other pathogens, notably viruses, subvert the normal functioning of cells. Some parasites merely rob the body of some of the nutrition it needs. Viruses must get inside cells before they can reproduce and do harm. Some bacteria also infect animals' cells: for example, the mycobacteria that cause tuberculosis in humans. Others kinds of bacteria, however, live in the spaces between cells—either within tissues or in body cavities.

Often, as with tuberculosis (TB), it is the body's own reaction to the presence of certain pathogens that results in disease. The mycobacteria that cause TB are difficult to kill; and once they have entered cells, the host's defense systems begin damaging its own cells and tissues in order to deal with the infection.

Body cells are constantly bathed in a complex blend of chemicals carried by the blood. Cells must be selective about which of these chemicals they let in or respond to, and which they keep out or ignore. Every body cell has surface receptors that act as gatekeepers, controlling the responses of the cell. Pathogens that infect body cells do so by binding to receptors on the surface and inducing the cell to pull them inside.

Often, disease begins when pathogens successfully bind to the surface of body cells. Successful binding occurs more often on mucosal surfaces, such as the lining of the nose or digestive tract, than on the skin. In the case of viral infection, this first binding allows an individual virus particle to enter and subvert the normal function of a cell. Information carried by the virus may then induce the host cell to manufacture more identical virus particles. However, some infections follow a more subtle and complex course.

Inflammation

Inflammation is a frequent symptom of many infectious diseases. In the case of human tuberculosis, mentioned above, a lot of the damage caused by the disease results because the human immune system attacks cells infected with the mycobacteria. The attack creates chronic (long-lasting) inflammation of the respiratory tract.

HOW INFECTION OCCURS

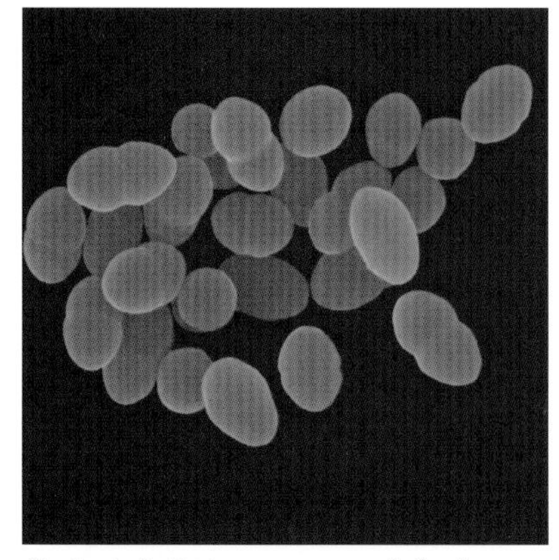

The bacteria that cause pneumonia (such as Streptocuccus pneumoniae*) are transmitted by small droplets of water vapor in the air. They cause inflammation of the throat and lungs.*

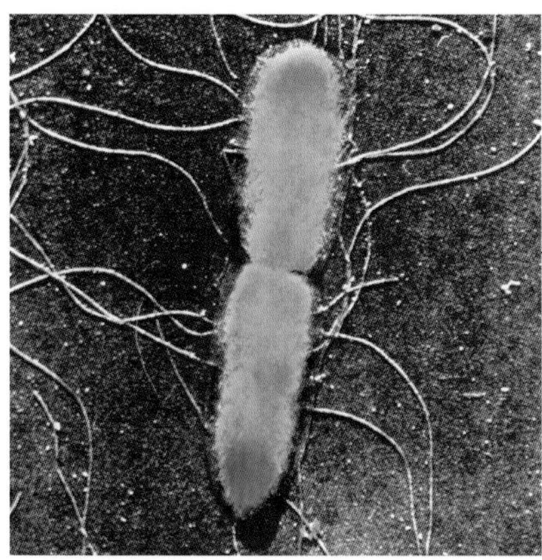

*The typhoid bacterium (*Salmonella typhosa*) enters the intestines through infected food or water and causes a serious generalized illness that is fatal in 25 percent of all cases.*

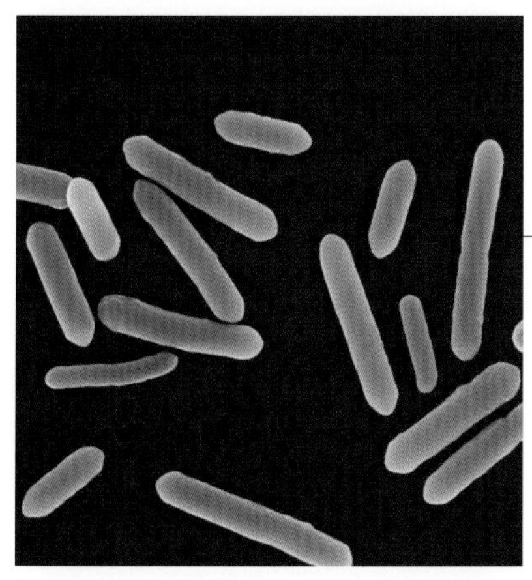

The bacterium that causes tetanus, or lockjaw, enters the body through any wound. Tetanus is characterized by muscle rigidity and spasms.

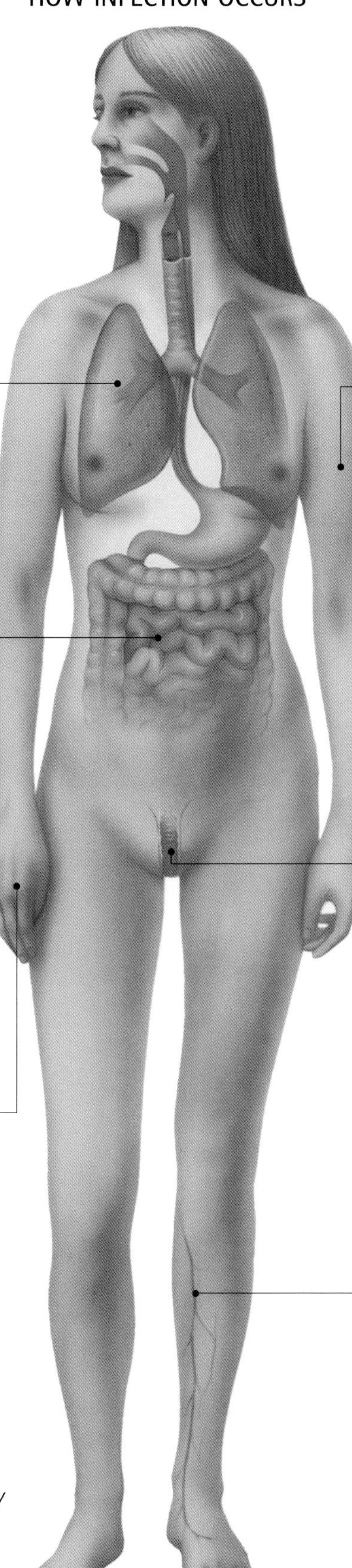

Several types of Rickettsia *bacteria cause the disease commonly known as typhus. These bacteria are carried by ticks, fleas, lice, and mites.*

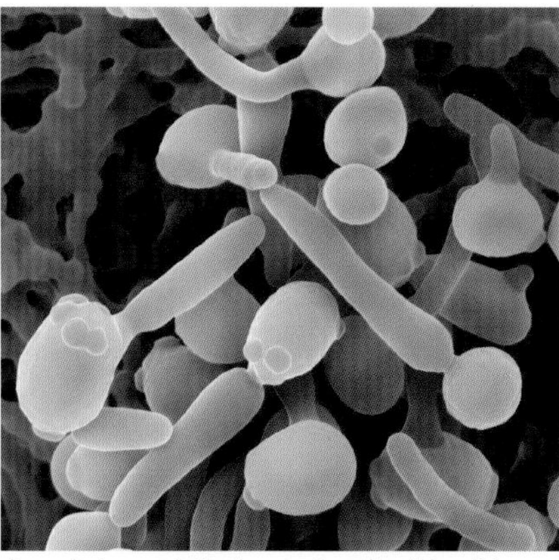

A genital yeast infection, such as Candida albicans*, is transmitted by sexual contact or by infected towels and in women induces a vaginal discharge.*

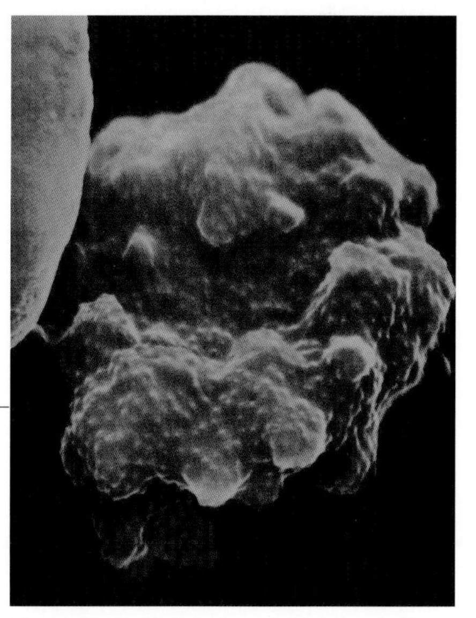

The protozoans that cause malaria are carried by mosquitoes. The protozoans are transmitted to humans by mosquito bites. This red blood cell has been infected by the malaria protozoan Plasmodium falciparum.

▲ SITES OF INFECTION

The site at which an infection enters the body depends upon the type of invading organism. Infections may result from bacteria, protozoans, viruses, fungi, or worms.

Some acute (short-lasting, but violent) inflammation results when the immune system deals successfully with the local attack of a pathogen. For example, when staphylococci bacteria enter a hair follicle in an animal's skin and begin to multiply, the animal will develop some soreness, reddening, and swelling as the inflammatory process locally increases blood supply. Lymph (a pale fluid containing the white blood cells of the immune system) also flows into the local tissue, which then swells. The inflammation is useful for the animal under attack because it identifies the enemy, increases the local forces to deal with the enemy, and keeps the defense forces well supplied. Sometimes, a little white "head" of pus may then accumulate as the next stage of inflammation. Pus is a concentration of white blood cells known as neutrophils. If the neutrophils successfully deal with the staphylococci—as they usually do—all is well and the inflammation ceases as the threat is defeated.

Through the long course of evolution, animals have developed different strategies to counter the many kinds of assault by pathogens. The pressure of natural selection on pathogens living in hostile environments (other organisms) that are trying to kill them has led to the evolution of a number of counterstrategies to avoid, evade, or disrupt animal immune responses. This kind of balance, where two kinds of organisms develop strategy and counterstrategy for exploitation or defense, is often described as an "evolutionary arms race." Similar arms races develop between predators and prey, parasites and hosts, and many other kinds of competing life-forms.

Types of immunity

Immune mechanisms can be thought of as falling into two main types: innate immunity and adaptive immunity. The idea behind this division is simple. Innate immunity comprises the parts of the immune system that exist whatever the host's experience of disease organisms has been in its lifetime. In contrast, adaptive immunity (also called acquired immunity) is influenced by the host's experience of pathogens (or of immunizations given to protect against pathogens).

Adaptive immunity is specific and "learned." Having experienced a certain type of pathogen once, the immune system of each host "remembers" it and reacts harder and faster to the same type of pathogen on a later occasion. This is why many diseases (such as chicken pox) are once-in-a-lifetime occurrences, and why vaccination can be such an effective defense.

The variable features of a host's adaptive immune responses derive from the activities of a group of cells known collectively as lymphocytes. Unlike any other body cells, lymphocytes are able to construct new working genes from component parts of the host's genome (the genome is the total genetic information contained in chromosomes, and is functionally divided into genes coding for proteins). By effectively creating new genes, the lymphocytes can produce potentially unique versions of proteins—although these are constrained by a basic general design. The products of these genes are highly specialized proteins known as antibodies and T-cell receptors. Both antibodies and T-cell receptors are involved in the recognition of foreign material (matter that originates outside the host's body, such as a pathogen or debris that has entered the body as a result of injury).

Adaptive immunity is thought to be restricted to vertebrates, but many more animals appear to have some level of innate immunity. Innate immunity does not involve lymphocytes. It is a combination of generally protective activities performed by a range of other cells and tissues. All complex organisms, including insects, annelids, mollusks, and many much simpler life-forms, possess innate immune responses.

The recognition of what is foreign to a host's body, and thus potentially a pathogen, is vital for immune responses to work. Just as important is knowing what belongs to the host's body, in other words what is "self," and thus not normally a threat. The immune response is not perfect: the innate part falls short of recognizing all pathogens, and the adaptive part sometimes starts to damage the host. This second phenomenon is called autoimmune disease, and rheumatoid arthritis is probably the best-known example.

IN FOCUS

Insect immunity

Insects possess innate, nonspecific immunity. They have no lymphocytes and they do not produce immunoglobulins. However, they do produce pathogen-destroying chemicals, for example, proteins called cecropins. Cecropins can successfully combat some, but not all, bacteria that infect insects by rupturing or "lysing" the bacterial cell membranes. A second class of insect immune chemicals, the attacins, act against a limited number of bacteria that tend to infect insects' alimentary canal. Many insect species also possess a type of lysozyme, an enzyme that lyses bacterial cells. All known insect immune responses are nonspecific and involve chemicals that are more or less successful at defeating insect pathogens.

Innate immune system

Innate immunity begins with an organism's surfaces: in mammals, these are mainly the skin and the mucus membranes. The outermost layer of human skin, for example, has considerable mechanical strength, is composed of dead cells, and is slightly acidic. This all helps make it an effective barrier. However, it is not impenetrable: hair follicles, sebaceous ducts, and sweat glands can all offer points of entry for a pathogen. The importance of skin as a first line of defense becomes very obvious when it is damaged. When large areas of skin areas are lost after serious burns, for example, the underlying tissue is highly vulnerable to infection and needs expert dressing to protect it. Similarly, a penetrating wound is often a source of bacterial infection.

Compared with skin, the moist, warm, mucous membranes are an attractive target for pathogens. Membranes' basic defense is the fluid coating of large sugar-rich mucin molecules on their surfaces. These molecules are kept moving by the beating of cilia in many areas (for example, in the trachea) or by peristalsis (in the intestines). The flow of mucus physically impedes the attachment of pathogens. Mucus also inhibits pathogen attachment chemically. Many viruses are able to bind onto complex sugars on the cell surface. Similar sugar residues (for example, Nacetyl galactosamine) in the mucus compete to bind the virus and bear it harmlessly away before it reaches the cell surface.

Interferon

In human tissues, one aspect of innate immunity is shared by almost all cells. Cells attacked by viruses produce a warning protein called interferon, which binds to receptors on nearby cells. As its name suggests, interferon acts as a signal to "interfere" with, or inhibit, viral replication in these cells. If a cell that has received an interferon signal subsequently becomes infected, it is already primed to "go slow" with regard to protein synthesis (viruses need protein synthesis to happen fast) and to make an enzyme to destroy ribonucleic acid (RNA), a chemical vital in gene expression. Without RNA, the virus cannot be replicated. Signaling molecules similar to interferon may also cause infected cells to "commit suicide."

EVOLUTION

Long history

Some innate immune responses have a very long history. The fact that all complex animals, from primates to slugs, possess receptors for the bacterial lipopolysaccharide (LPS) molecule implies that the receptor must have evolved before the common ancestors of higher animals diverged, more than 500 million years ago. The complement cascade is also very ancient; a version is known to have evolved in echinoderms (sea urchins) about 500 million years ago.

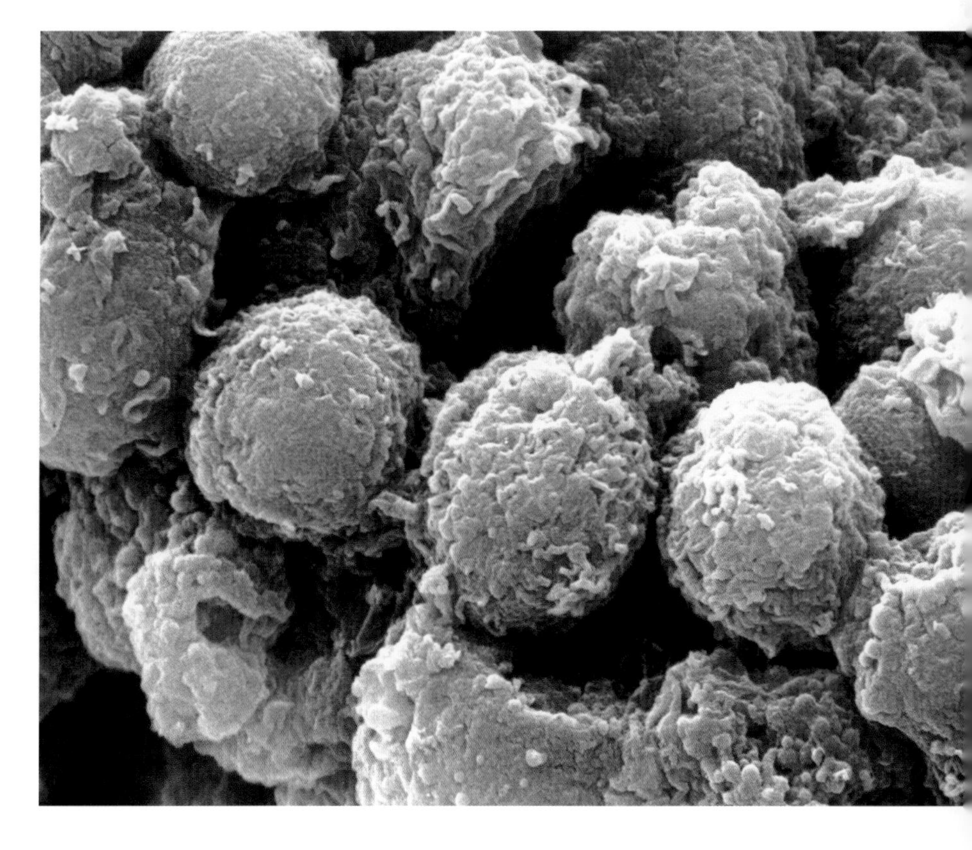

▼ *These monocytes are white blood cells that have formed in the bone marrow. They are phagocytic, engulfing and digesting invading pathogens.*

CLOSE-UP

All systems go!

The action of antibodies is normally considered under the heading of adaptive immunity, but antibodies also have functions in the innate immune system. Scientists tend to divide the immune system into innate and adaptive mechanisms for purposes of understanding, but in reality when it comes to dealing with infection, everything works together. Antibody molecules can refine and focus not just the binding of molecules in the complement cascade but also the activities of phagocytic cells and of those leukocytes such as eosinophils that kill without engulfing. So, despite their origin in the adaptive immune system, antibodies do not generally achieve a separate solution to infectious disease. Instead, they are part of an integrated assault on pathogens from both innate and adaptive systems.

Cell suicide is called apoptosis. This emergency measure can be a useful weapon against some pathogens: dead cells are useless to invading viruses and thus tend to stop the infection from spreading.

Inevitably, infections will sometimes start to take hold. A huge range of innate defenses can begin to operate within an organism that is under threat, but a great deal depends upon their recognizing the pathogen involved. A host that does not know an enemy has entered cannot take steps to remove it.

Animals have receptors on cell surfaces and on free molecules in body fluids to immediately recognize various molecular patterns or shapes that are characteristic of pathogen surfaces. Bacteria, fungi, and viruses all have these distinctive surface patterns, and all complex animals have receptors that can recognize at least some of them. For example, bacteria typically have a coat containing a chemical called lipopolysaccharide (LPS). LPS possesses a molecular pattern that sets alarm bells ringing in the innate immune systems of everything from humans to beetles or snails. When a host's defensive cells bind LPS on their surfaces, they are activated to start a sequence of defensive measures.

Toll-like receptors

Other important molecules whose patterns are recognized by the innate immune systems of most animals include the protein flagellin (so called because it is found in the whiplike flagella that bacteria use for movement) and various bacterial forms of nucleic acids. Like LPS, all these pattern molecules are recognized

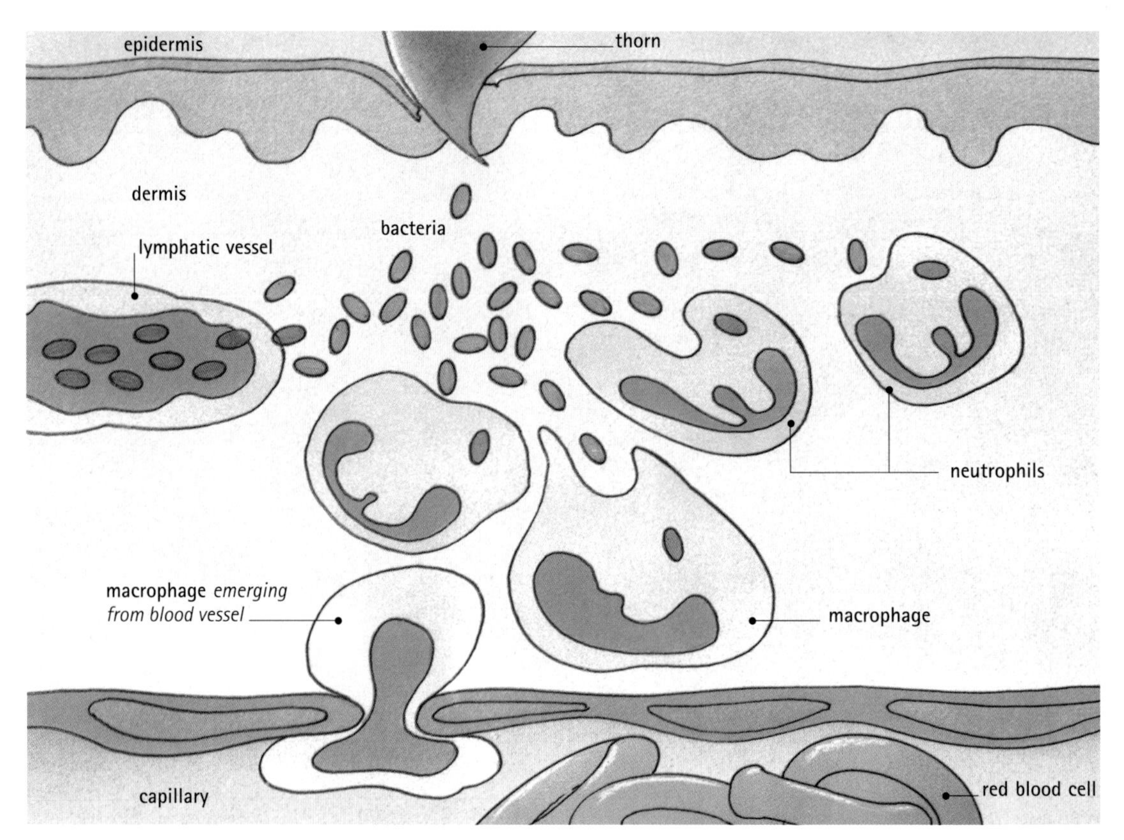

▶ INNATE IMMUNITY

In the natural immune system, invading pathogens such as bacteria are ingested and destroyed by cells called neutrophils and macrophages.

by a family of receptors called toll-like receptors, or TLRs, which are found on animals' innate immune cells. In mammals, TLRs are well represented on types of cells called dendritic cells, which are embedded throughout the animal's tissues, especially near body surfaces. Although the role of dendritic cells is not fully understood, it seems that when the TLRs recognize a warning sign in the form of a particular molecular pattern, the dendritic cell sends out a signal in the form of a type of protein called a cytokine. Cytokine release coincides with the start of both inflammation and immunity. Dendritic cells act as sentry guards, continuously sampling their local environment and poised to trigger local defense mechanisms as soon as a pathogen is detected in the vicinity.

The complement cascade

Dendritic cells may not provide the only warning signals. Not all pattern-recognizing receptors are attached to cell surfaces. Some such as the mannose binding lectin (MBL), which recognizes the bacterial surface sugar mannose, are soluble and exist free in the host's body fluids. MBL sticks to mannose and begins an immune reaction called a "complement cascade." The reaction leads to the formation of small molecules called anaphylatoxins, which rapidly diffuse away from the site and act in several ways. As with the cytokines produced by dendritic cells, anaphylatoxin release can be associated with an accumulation of neutrophils (pus cells), and changes to the local blood supply, bringing more defenses to the infected area. The anaphylatoxins also induce capillaries to leak plasma into the affected tissue, making it swell.

The complement cascade (so called because it supports, or complements, the action of antibodies) is an immune reaction involving about 30 proteins that interact and modify one another. Initially these interactions take the form of a positive feedback loop: that is, the initial response is repeatedly magnified and can rapidly lead to a major reaction. In addition to the release of anaphylatoxins, the cascade also leads to the formation of a "membrane attack complex," which helps destroy pathogens by punching holes in their cell walls. It also triggers the deposition of a product called C3b on the surface of pathogens. This makes the particle much more likely to be taken up, and therefore destroyed, by a phagocytic cell. This process is called opsonization.

Leukocytes

In addition to neutrophils, the cells of the innate immune system also include several other types of white blood cells, or leukocytes, which may work separately or together to kill an invader. The protective roles of mast cells, basophils, and eosinophils are probably connected with killing large pathogens, such as parasitic helminth worms. Neutrophils are phagocytic leukocytes (named for the Greek *phagein*, meaning "to eat"). They patrol the body tissues, on the lookout for pathogens to engulf and consume. Macrophages ("big eaters") are cells that routinely collect and consume cell debris but also engulf any pathogens they may find. Neutrophils and macrophages are capable of killing most bacteria, which they engulf in a membrane-bound organelle called a phagosome. In the cell's cytoplasm, organelles called lysosomes fuse with the phagosome, producing a phagolysosome. The lysosomes deposit powerful chemicals into the space containing the bacteria. The bacteria are attacked by these compounds, which include free radicals and destructive enzymes. Some pathogens can withstand this kind of treatment, but it is generally effective at killing bacteria.

Leukocytes also contain peptides (molecules that are structurally similar to proteins but smaller) called defensins, which bind into and damage the cell walls of bacteria. Defensins are certainly important, but their role in the innate immune system is not fully understood.

IN FOCUS

Crybaby

Tears are an innate immune mechanism that protects the surface of the mammalian eye by creating a flow to impede pathogen binding. Tears also contain a good antibacterial agent called lysozyme. This acts in a way similar to some antibiotic medicines, killing bacteria by attacking their cell walls.

CONNECTIONS

COMPARE the immune system of a ***CRAB*** with that of a ***HUMAN***. Crabs are probably just as able as a human to fight off pathogen attacks, but they defend themselves differently. Human blood contains about 30 different proteins that can release anaphylatoxins to help destroy disease-causing pathogens. Horseshoe crabs rely on just two proteins, limulin and alpha 2-macroglobulin, to accomplish this task.

Adaptive immune system

Antibodies (sometimes known collectively by the singular term "antibody") are the immune system molecules that most people have heard of. Antibody is produced by one of the great classes of lymphocytes (called B lymphocytes, or "B cells"). B cells not only secrete antibody but also bear antibody molecules on their surface membranes. These surface immunoglobulins, or sIgs, are representative of the specific antibody the B cell would secrete into the blood serum (the liquid component of blood, in which blood cells are suspended) if the cell was turned on. Antibody molecules are proteins with some sugar residues attached.

The most common kind of antibody, known as immunoglobulin G (IgG), has a basic shape similar to the letter Y. The base or foot of the Y (known as the Fc region) is where an antibody can dock to and interact with a host cell, or interact with proteins of the complement system. The tips of the arms of the Y (the molecule's "hands") are where the antibody binds with its target or "antigen," usually a foreign material. The precise structure of the "hands" varies greatly, thus allowing different antibody molecules to bind specifically to different antigens, or to different parts of the same antigen. The scientific name for the molecule's "hands" is complementarity determining regions (CDRs). Roughly, their

CLOSE-UP

Types of antibodies

The basic Y antibody is called IgG (Ig is short for immunoglobulin). There are various other, less common antibody types. For example, IgM looks like five Y's arranged in a star with their bases joined together, and IgA can resemble a single Y or two Y's joined tail to tail. Antibody molecules are found in the blood serum and lymph, and so bathes the tissue spaces. Huge amounts of IgA are made and poured onto mucous surfaces such as those of the intestines.

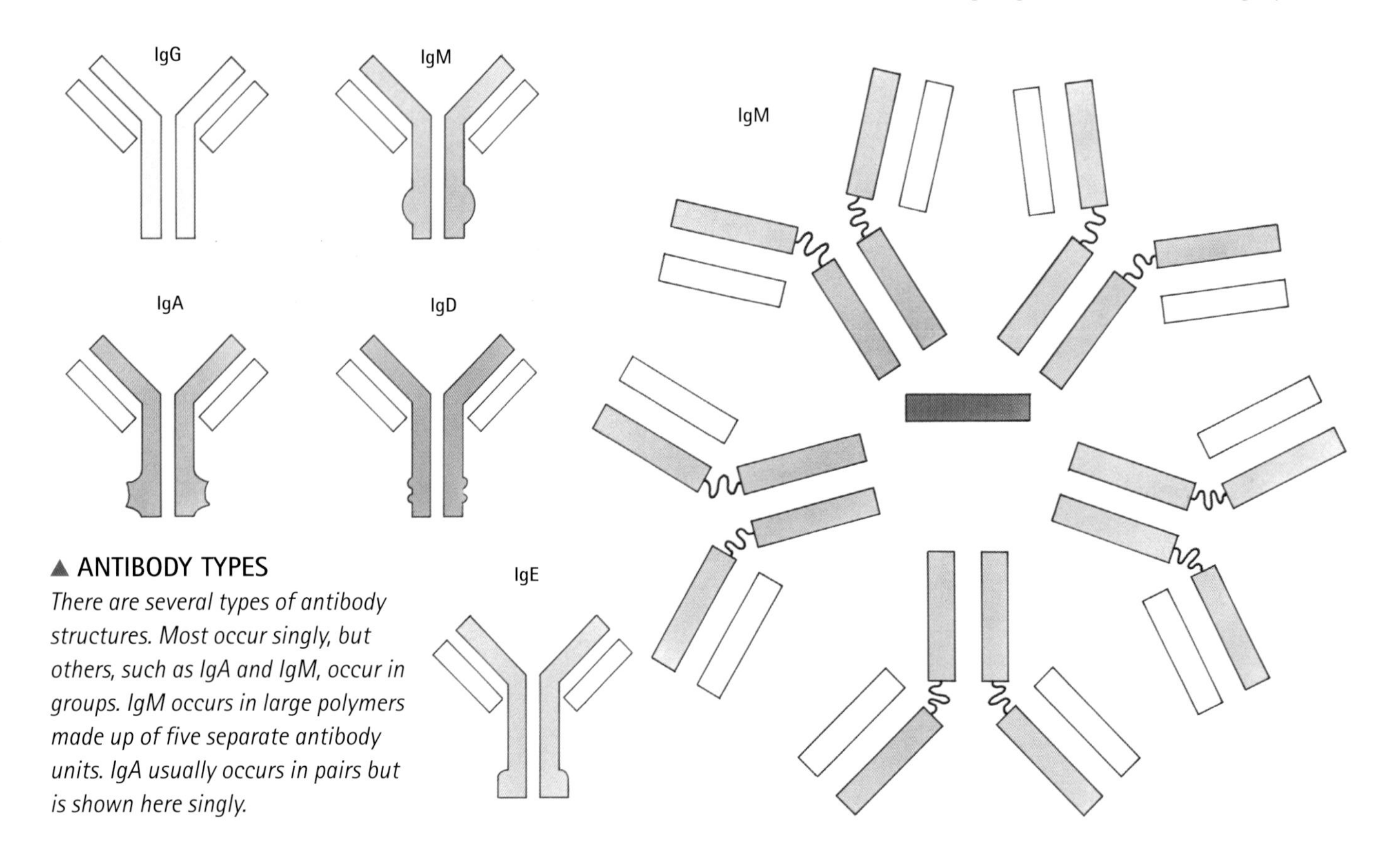

▲ ANTIBODY TYPES

There are several types of antibody structures. Most occur singly, but others, such as IgA and IgM, occur in groups. IgM occurs in large polymers made up of five separate antibody units. IgA usually occurs in pairs but is shown here singly.

molecular shapes are "complementary" to the molecular shape of the antigen concerned, so they fit together in the same precise way as a lock and key. In fact, think of those keys where you can see a flat surface with hollows in it; inside the lock is a flat surface but with bumps that complement the hollows. Antibody-antigen binding works in a comparable way.

The function of antibodies

The simplest role of antibody molecules is neutralization or blocking. A microbe, especially a virus, is neutralized if antibody binds to the molecules it would otherwise use to latch onto and invade a cell. The antibody acts as a simple blocker, preventing the virus from linking up with a cell membrane. Antibody molecules can also neutralize toxins. Many pathogenic bacteria produce chemicals that are poisonous to host cells. An antibody molecule that binds to a biologically essential part of the toxin can render it nontoxic.

Another important antibody activity is agglutination. Agglutination is the process by which an antibody molecule links two other molecules or cells, forming a bridge between them. They are in effect stuck together with the antibody as glue. IgM, with its fivefold starlike array of Y-shapes, can cross-link as many as 10 antigen sites. Pathogens or chemicals held together like this may be disabled.

A third crucial role of antibody molecules is opsonization. Any antigen that has been bound by an antibody molecule, regardless of whether or not it has also been agglutinated or neutralized, is marked out for destruction by phagocytic cells. In this respect antibody is working somewhat like C3b. The opsonizing activity works because many phagocytic cells have receptors on their surfaces for the foot, or Fc region, of the antibody molecule. These receptors are thus known as Fc receptors; this is where the antibodies "dock."

So in general, antibodies operate separately from the lymphocyte cell that produces them, attacking pathogens in the body fluids, or docking on the surface of a different cell type, such as a phagocyte, and influencing its activity.

Phagocytes dispose of pathogens by engulfing them, but other types of cell have different means of attack. Some cells kill pathogens by shooting out toxic contents onto a target such as a pathogen or an infected cell. Usually the toxins act by rupturing or "lysing" the target cell. In this instance the antibody molecules direct the attack by binding to the target and linking it to the killer call. For example, if one of your own cells was infected by a virus that was replicating, then killing the infected cell might be the best way to prevent many neighboring cells from becoming infected. This carefully targeted poisoning of marked cells is called antibody-dependent cell-mediated cytotoxicity (ADCC). There are several sets of killer cells that work in this way. It is a highly effective system, but there can be unfortunate complications—a form of allergy.

A further very important role of antibody is that it can focus innate immune responses such as those of the complement cascade system, just as it focuses ADCC. When an antibody

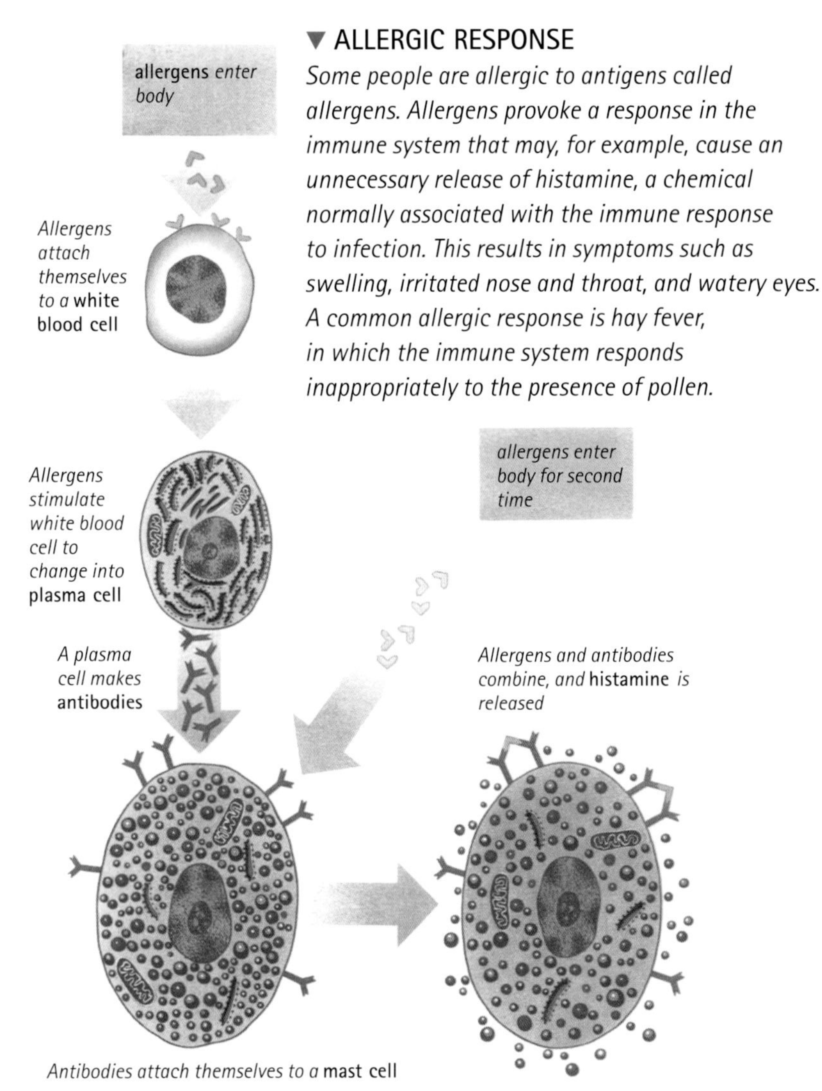

▼ **ALLERGIC RESPONSE**
Some people are allergic to antigens called allergens. Allergens provoke a response in the immune system that may, for example, cause an unnecessary release of histamine, a chemical normally associated with the immune response to infection. This results in symptoms such as swelling, irritated nose and throat, and watery eyes. A common allergic response is hay fever, in which the immune system responds inappropriately to the presence of pollen.

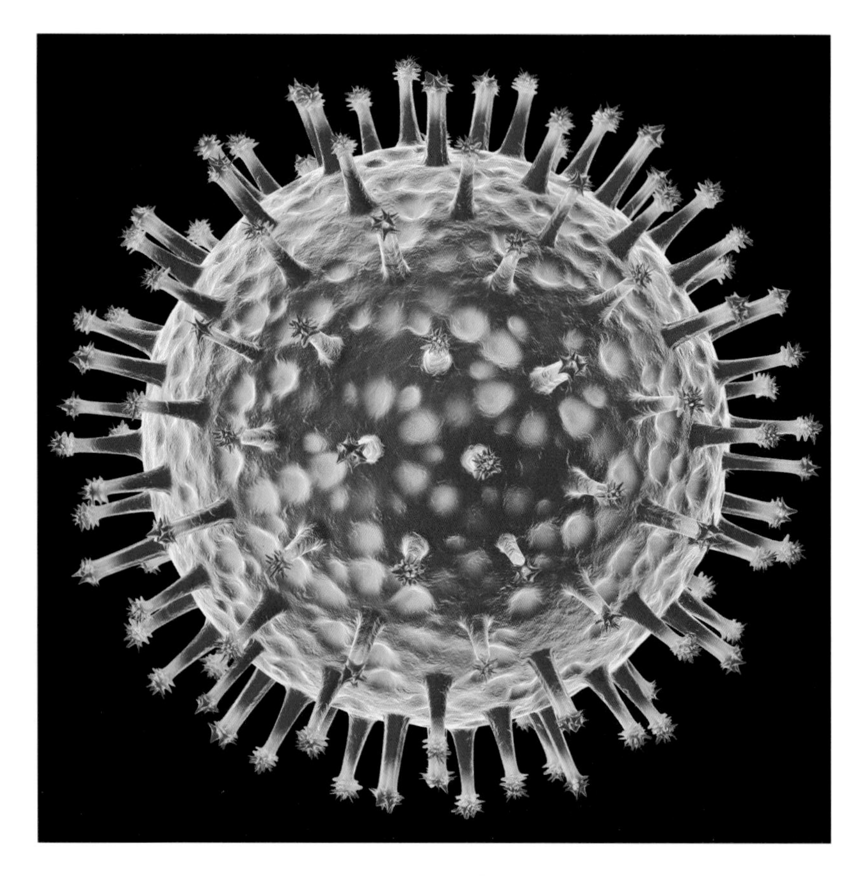

molecule binds to its antigen, a subtle change occurs in the shape of the foot or Fc region. One of the complement proteins is attracted to this changed shape, and the interaction triggers the cascade system. The complement cascade acts hard and fast. Using its positive feedback loop, it brings all the vigor of the innate response to bear directly on the culprit organism, summoning an army of neutrophils and initiating other physiological responses. In effect, the innate immune system is a blunt weapon that can be honed to deadly accuracy by the precise ability of the adaptive system to identify trouble. It is a beautiful example of cooperation between systems.

◀ *Viruses, such as this bird flu virus, can reproduce only inside a cell. They gain entry to the cell by binding to surface receptors and inducing the cell to pull them inside. Once inside, they alter the cell's processes, causing it to produce more virus particles. The cell, however, also produces a protein called interferon, which inhibits the further replication of the virus in surrounding cells.*

▼ **Antibody structure** *An antibody is composed of heavy and light chains of molecules held together by disulfide bridges. Each antibody unit has two antigen binding sites. These are the regions which are most structurally diverse and which evolve by processes of mutation and selection. Thus the adaptive immune system is able to respond to new pathogens and, in most cases, defeat them.*

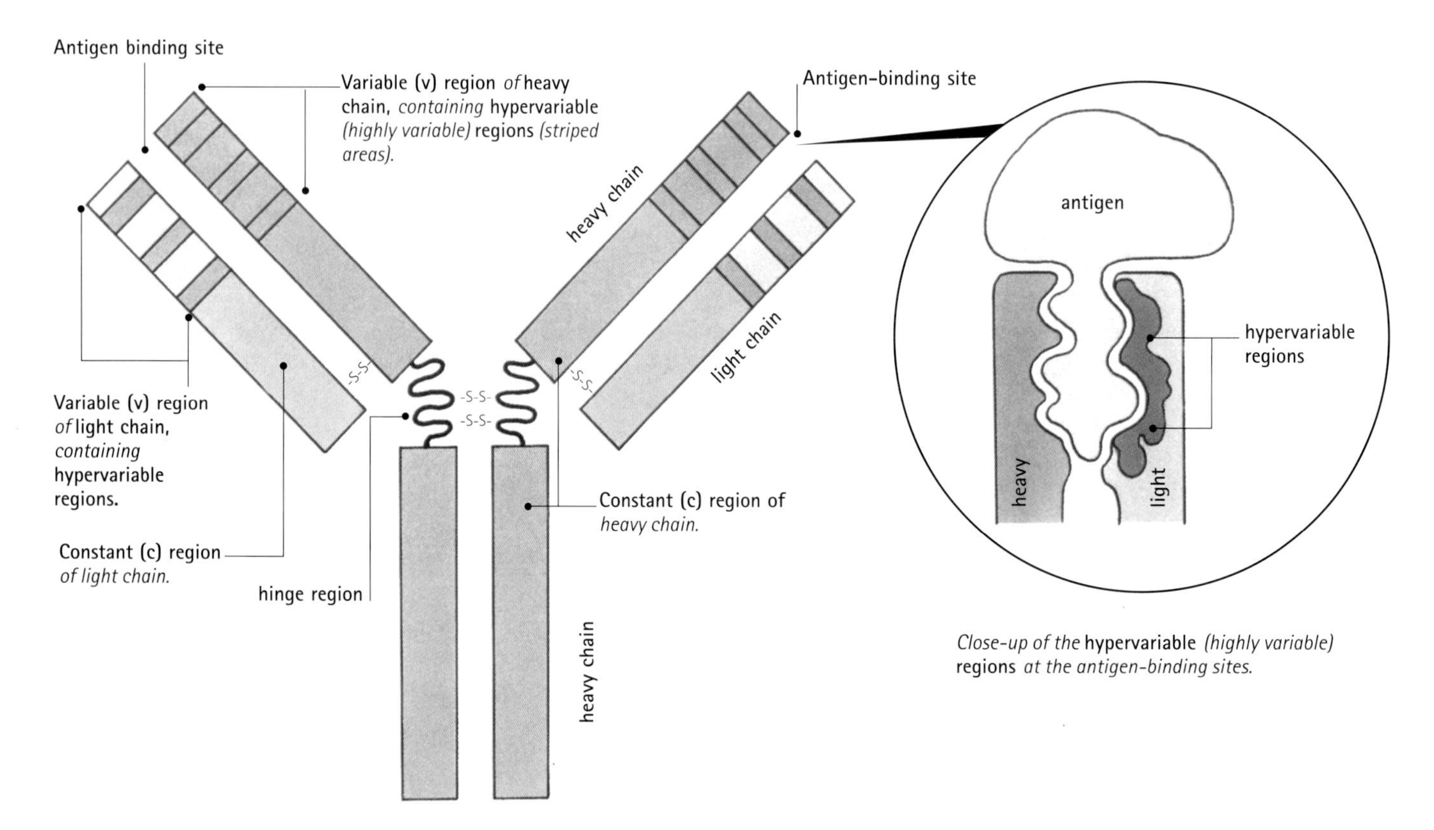

Close-up of the hypervariable *(highly variable)* regions *at the antigen-binding sites.*

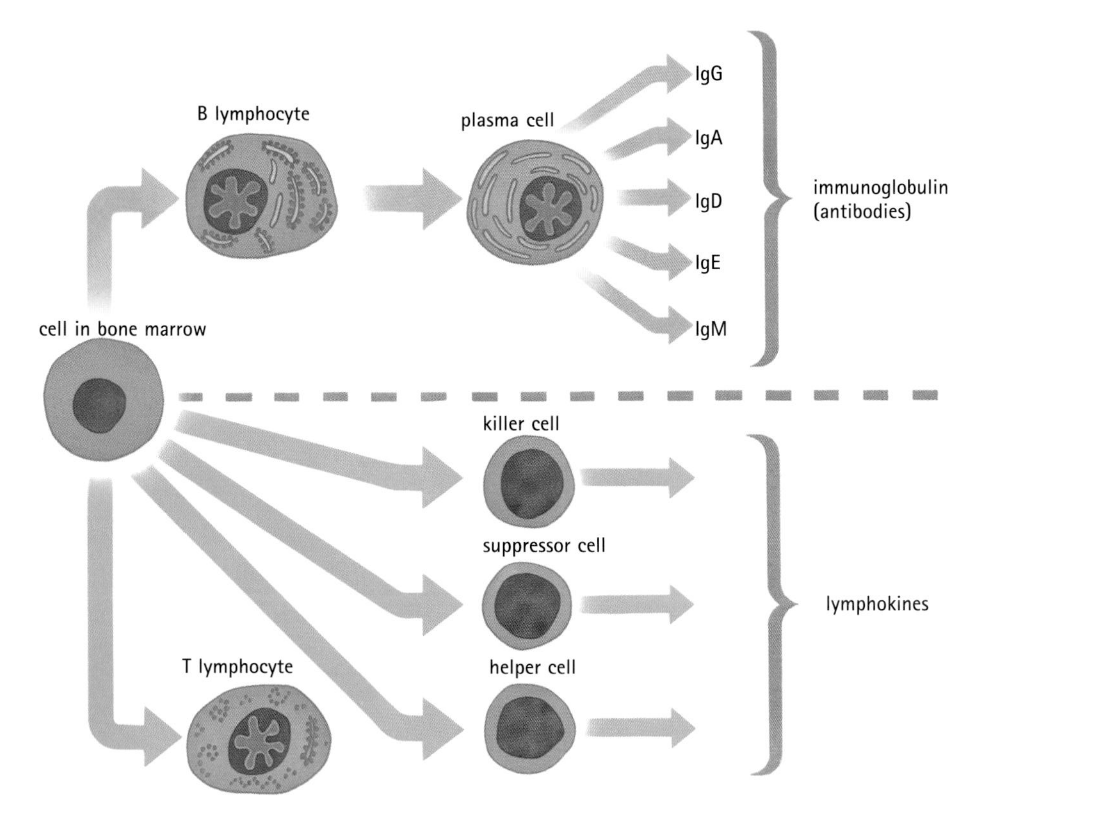

▲ IMMUNE CELL DIVERSITY

A bone marrow stem cell may divide into a B lymphocyte (B cell) or a T lymphocyte (T cell). These two types of immune cell differentiate further into specialized immune cells with specific functions in fighting disease.

CONNECTIONS

COMPARE the immune system of a vertebrate such as a ***HUMAN*** with that of an invertebrate such as a ***STARFISH***. In addition to the innate immune system, a human has an adaptive immune system based on antibodies; but starfish, spiders, insects, and many other invertebrates have an immune system that is not based on antibodies.

Allergies

Mast cells are part of the immune response that probably evolved in response to large parasitic worms, whose cuticles (external surfaces) are very resistant to attack. Mast cells and eosinophils attack using the antibody-dependent cell-mediated cytotoxicity (ADCC) described earlier. Mast cells bear a form of antibody called IgEs. In parts of the world where parasitic worm infestation is routine, this manifestation of the immune system is fundamentally useful. Unfortunately IgEs are also frequently made in response to a set of antigens that are not directly pathogenic. We call these antigens allergens. Pollen, house dust mites, and cat dander (minute scales from skin or hair) are examples of common allergens. When an allergen cross-links an IgE molecule on a mast cell surface to another IgE molecule on the mast cell, the cell releases a cocktail of inflammatory molecules into the location, including the protein histamine, and an allergic episode is started. Because of the nature of the allergens, the site of first contact is likely to be an exposed body surface or airway—hence the itching, runny eyes and nose, and irritated throat experienced by allergy sufferers.

CLOSE-UP

Surface immunoglobulins

B-cell surface immunoglobulin (sIg) is a special case. This is a type of antibody found on the surface of B lymphocyte cells. It represents the specificity of antibody that B cells produce if they develop all the way into secreting plasma cells. sIg is crucial to the fate of the B cell. If sIg reacts with "self," the B cell will be programmed to die by an interaction with the cells that surround it in the marrow before it goes out into the circulation. If the cell passes this test (by not recognizing anything) the next job of sIg will be to stimulate the cell to divide, if and when antigen that matches the sIg is encountered, during an infection.

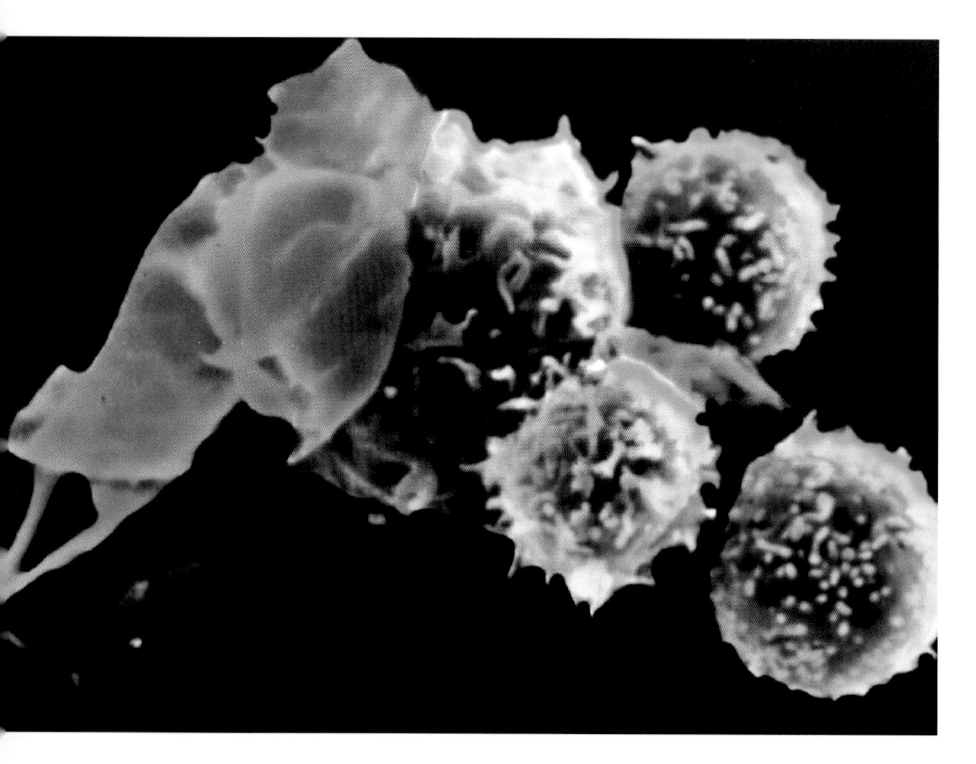

▲ *A colored scanning electron micrograph (SEM) of four killer T lymphocytes (shaded blue) attacking a cancer cell (shaded red). The killer T lymphocytes recognize the cancer cell by its surface antigens.*

Allergies can have other origins apart from IgE and mast cells, but this form, with its rapid onset, is referred to as immediate hypersensitivity. Its most severe manifestation is anaphylactic shock, which can be quickly fatal. The allergen in this reaction is likely to have been injected: for example, an insect venom.

HIV and AIDS

When the human immunodeficiency virus (HIV) enters target cells, it injects its genetic material into the cell nucleus, where the viral genes are spliced directly into the cell's own DNA. Thereafter, every time the cell replicates its own genome and divides, it also copies HIV.

For 7 to 10 years after infection, the body produces helper T cells, which mobilize cytotoxic T cells to fight the infection. No one knows why, but infected people have no symptoms during these early years. The immune system battles ferociously, but in most cases, without the assistance of medication it ultimately loses.

At some point, HIV mutates in such a way that infected dendritic cells pass HIV on to helper T cells in the lymph nodes. HIV kills the helper T cells. The body boosts its production of helper T cells in efforts to activate B cells and killer T cells, but the more helper T cells are produced, the worse the infection becomes and the more helper T cells are killed by the virus. When the level of helper T cells in the blood falls below a certain level, the infected person's immune system is very seriously impaired, and the person has AIDS (acquired immune deficiency syndrome).

Normal immune system

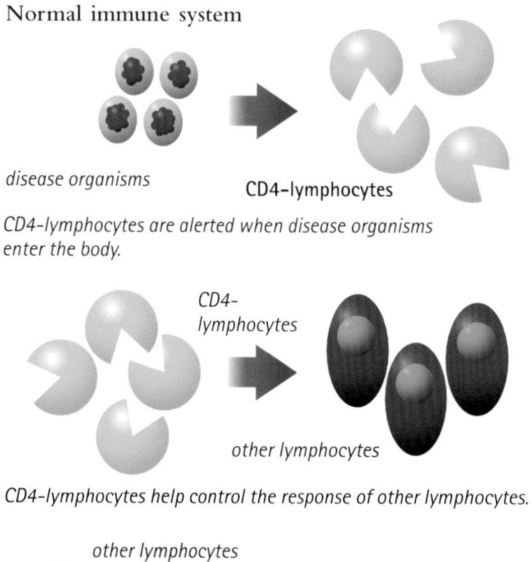

CD4-lymphocytes are alerted when disease organisms enter the body.

CD4-lymphocytes help control the response of other lymphocytes.

other lymphocytes

disease organisms

The other lymphocytes destroy the disease organisms.

Immune system in person with AIDS

HIV

CD4-lymphocytes

HIV multiplies inside the CD4-lymphocytes and may destroy them.

disease organisms

destroyed CD4-lymphocytes

The body's immune responses may fail when disease organisms invade because the CD4-lymphocytes have been destroyed.

disease organisms

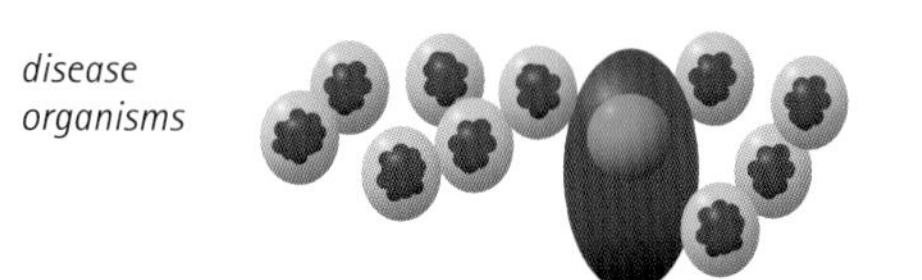

The disease organisms may overwhelm the immune system.

► HIV

HIV destroys the CD4-lymphocytes, which help to control the action of other lymphocytes. Thus the immune system is impaired and may not be able to rid the body of some disease-causing organisms, leading to a combination of illnesses diagnosed as AIDS.

Major histocompatability complex

One apparent drawback of a highly developed adaptive immune system is the difficulty it poses for tissue and organ transplantation. For example, if an organ like a kidney is randomly taken from one individual and implanted into another individual it will have little chance of surviving, because it will be recognized as foreign by the recipient's immune system and rejected. Through the study of transplantation success and failure (for example, with skin grafts between mice), scientists were able to understand tissue compatibility, which is also called histocompatibility.

Among the many different molecules present on the surface of cells are proteins of the major histocompatibility complex (MHC). These proteins act as advertisements, showing what is happening inside the cell. They can signal "business as usual," or indicate that something different is happening within. There are two different types of advertising molecules: MHC class I and MHC class II molecules. MHC class I molecules, which are displayed on the surface of most cells, contain peptides that represent the kinds of compounds the cell is making. MHC class II molecules, on the other hand, display peptides that originate from what the cell is degrading. Not all cells bear MHC class II molecules—generally these advertisements appear on specialized cells associated with immune function, such as those in the lymph nodes. Usually, both the manufactured peptides advertised by MHC class I molecules and those being broken down and advertised by MHC class II molecules will be bits of "self" protein. Their manufacture and disassembly are part of normal cell activity. Thus when MHC I and MHC II molecules both display "self" peptides, the signal is "business as usual," and the cells attract no attention.

However, if a tissue cell is infected by a virus, its protein synthesis mechanisms are partly taken over, and peptides that are definitely "nonself" will be produced and displayed on the surface by MHC class I molecules. This attracts the attention of cytotoxic T cells, whose receptors (TCRs) are "tuned in" to changes in MHC class I structure. On detecting foreign peptides, the T cells turn into potential killers and the cell making virus is likely to be destroyed. The "tuning in" of cytotoxic T cells to MHC class I depends on another cell surface molecule called CD8. MHC class I molecules can alert cytotoxic T cells to the presence of antigens in almost any cell in an animal's body.

If a cell bearing MHC class II molecule signals the destruction of a foreign material within, the immune response is rather more complicated. MHC class II molecules are recognized not by cytotoxic T cells but by helper T cells. These have the molecule CD4 on their surfaces. Helper T cells will not kill the cell carrying the foreign MHC class II advertisement, but having been alerted, they start to divide and produce molecules that will stimulate neighboring B cells into producing more antibody. B cells are likely to have begun gathering in the area already, having separately recognized the foreign antigen. Within a lymph node, any foreign-derived material arriving may be detected in two ways: directly by B cells and indirectly as pieces of peptide by helper T cells.

▼ This baby was born with a serious heart defect and required a heart transplant. To reduce the risk that a transplant organ will be rejected by the immune system, the organ must be "typed" to ensure that its antigens are as similar as possible to those of the recipient. In addition, drugs that suppress the immune system must be administered.

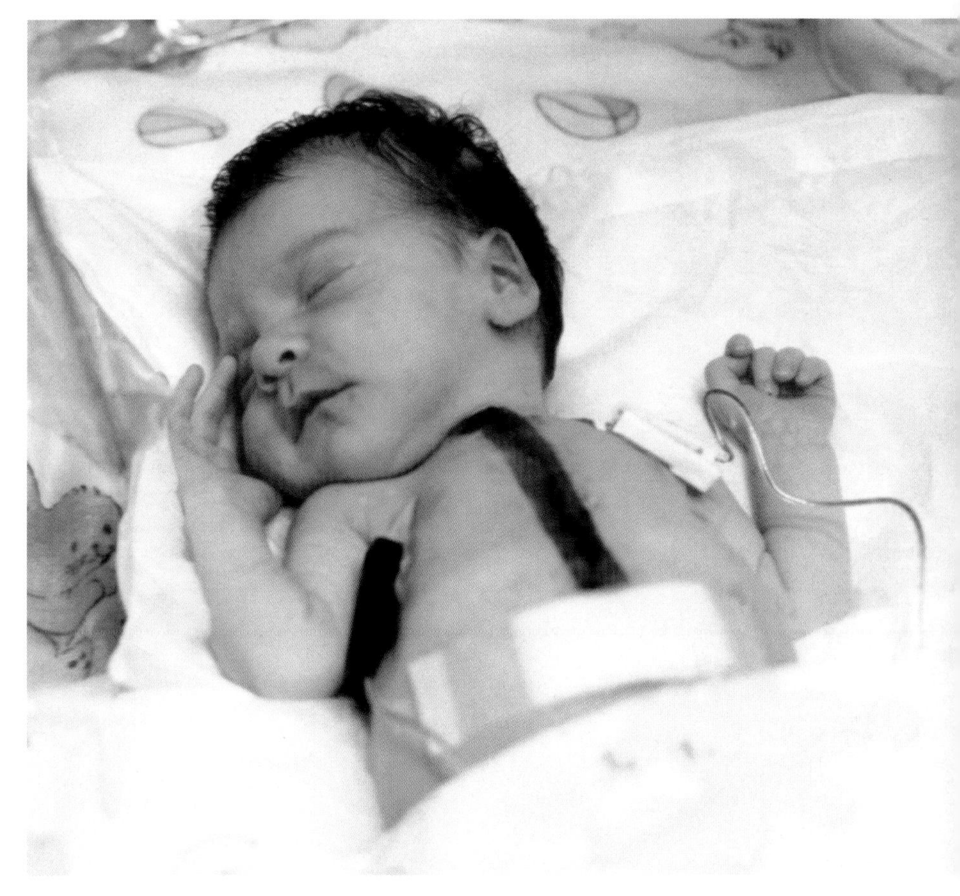

Lymphatic system

The lymphatic system consists of a series of ducts, or lymphatics, connected to numerous small masses of tissue called lymph nodes. The function of the lymphatic system is to collect, sample, and recycle body fluid. As blood circulates, its plasma—clear, colorless liquid containing oxygen, proteins, glucose, and other nutrients—seeps through capillary walls to refresh the extracellular fluid between cells. Over time, plasma can either seep back into the bloodstream or be drained by the lymphatic system. The lymphatic system carries this fluid (now called lymph) through the lymphatics until it reaches one of the lymph nodes. Lymph nodes filter material from the lymph and test it for certain constituents. Filtered material includes the degraded remains of the host's own dead cells, but the lymph may also include foreign matter draining from the site of an infection. Lymphocytes (T cells and B cells) gather in different parts of the lymph nodes, and when receptors on specific lymphocytes bind to the foreign antigens, an immune response to the infection is launched. Almost all of this activity depends on major histocompatability complex molecules. The activity in a lymph node during infection can be enough to make the node become enlarged and tender. For example, a person may have swollen "glands" (lymph nodes) in his or her neck when suffering a streptococcal throat infection. Lymphatic vessels are located in most parts of the body, but slightly different structures serve in the drainage of mucosal surfaces such as the gut and bronchi.

Origin of lymphocytes

Lymphocytes are "born" in bone marrow, the soft tissue that fills many of our long bones. Bone marrow contains stem cells that are constantly dividing to produce immature lymphocytes, which are released into the bloodstream. Immature lymphocytes do not carry specific receptors but develop such receptors when they mature. B lymphocytes mature in a range of sites including the bone

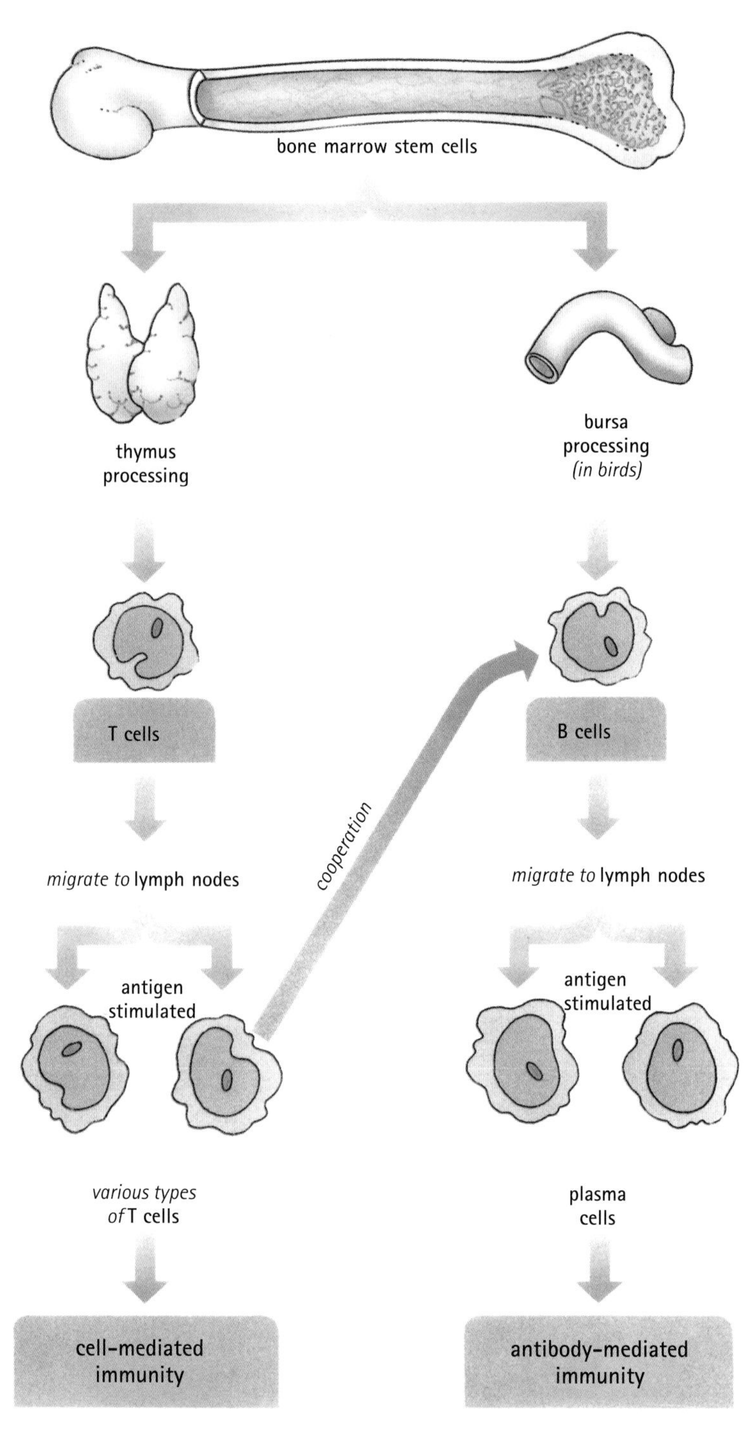

▲ PRODUCTION OF LYMPHOCYTES

B cells and T cells originate in the bone marrow, where they are produced by stem cells. B cells mature in the bone marrow, except in birds, where they mature in a lymphatic organ called the bursa of Fabricius. T cells pass to the thymus, where they mature into helper T cells and cytotoxic T cells.

LYMPHATIC SYSTEM

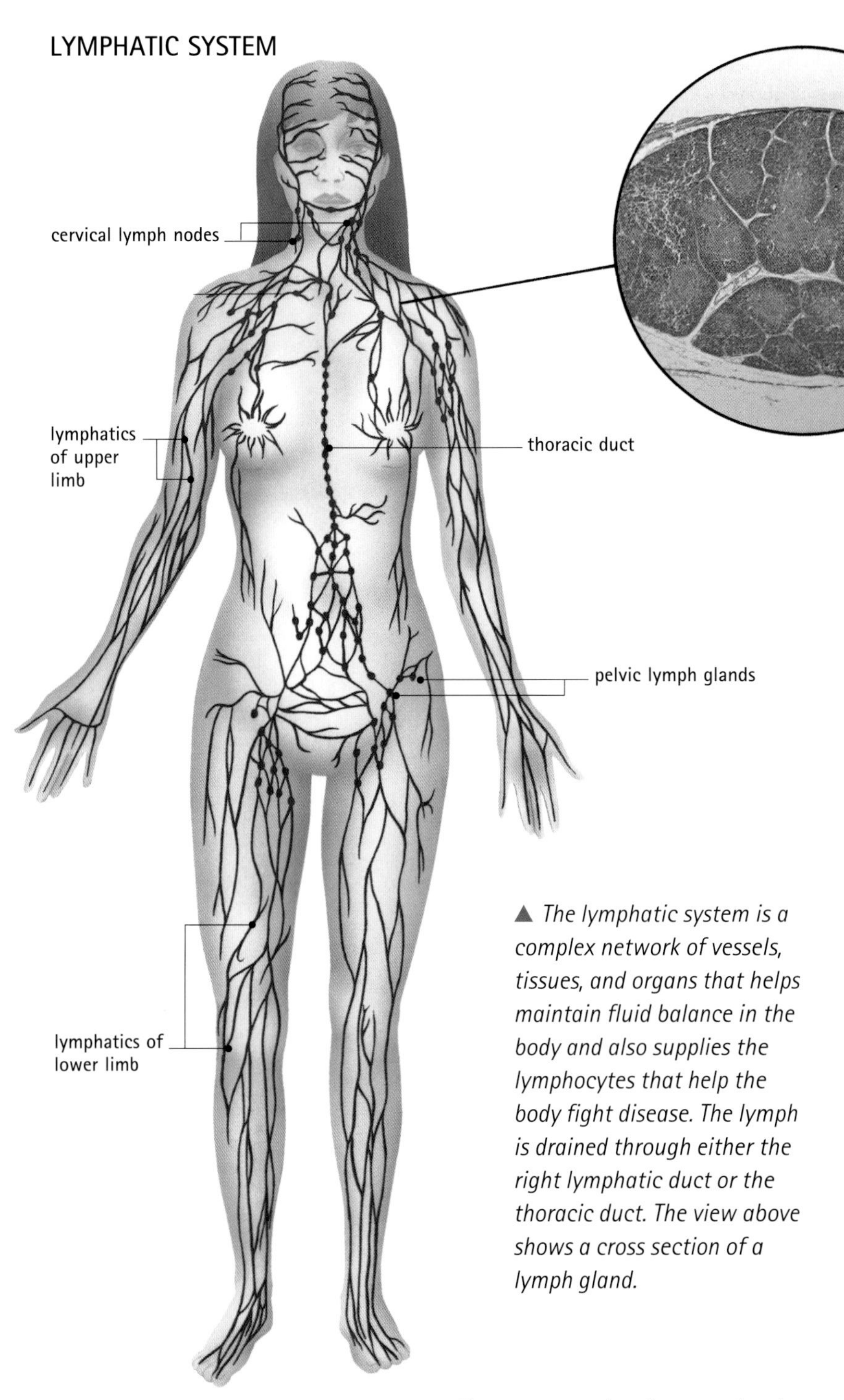

▲ *The lymphatic system is a complex network of vessels, tissues, and organs that helps maintain fluid balance in the body and also supplies the lymphocytes that help the body fight disease. The lymph is drained through either the right lymphatic duct or the thoracic duct. The view above shows a cross section of a lymph gland.*

marrow; T cells mature in the thymus gland—a part of the lymphatic system situated in the neck or chest. The human thymus gland is located in the chest, just behind the breastbone. Infants have a large thymus, but the gland dwindles after puberty. Within the thymus, T cells multiply repeatedly as they move through the network of thymus cells. T cells present in the thymus are called thymocytes. During this process, the thymocytes differentiate into helper T cells and cytotoxic T cells and acquire receptors that enable them to detect specific antigens. Any thymocytes having receptors that could react with "self" material are eliminated. So rigorous is the testing of thymocytes that only about 5 to 10 percent of all thymocytes created become full-fledged T lymphocytes and move on to life in the immune system.

Generation of diversity

There is a fundamental similarity in the way that the huge numbers of diverse new receptor molecules are made in both B and T lymphocytes. This diversity is achieved through creating new working genes by recombining segments from several fairly limited sets of component gene pieces—a bit like a jigsaw puzzle in which the pieces can be put together different ways to make different pictures.

Why do we need an adaptive immune system at all? We have innate immunity to a number of pathogens recognized by the toll-like receptors (TLRs) of the dendritic cells. In theory the number of different TLRs could be extended to include every sort of foreign molecule. In reality such a system would be colossally inefficient. Not only would the number of genes required to provide full innate protection be unthinkably huge, but the majority of the receptors would never be used—even the most unfortunate individual is likely to encounter only a fraction of the potential pathogens that exist in the world. Not only would such a system require an impossibly large genome; only one new, unrecognizable pathogen could defeat the whole system.

Consequently, the system that has evolved in vertebrates (with a primitive form in the jawless hagfish, and a more advanced form in sharks) is an adaptive system in which cells are able to make a large variety of receptors by recombining a relatively small number of gene segments—just as we can build an enormous number of words from a set of just 26 letters. These cells also induce mutation in limited, but functionally crucial, regions of the newly assembled genes. The range of potential

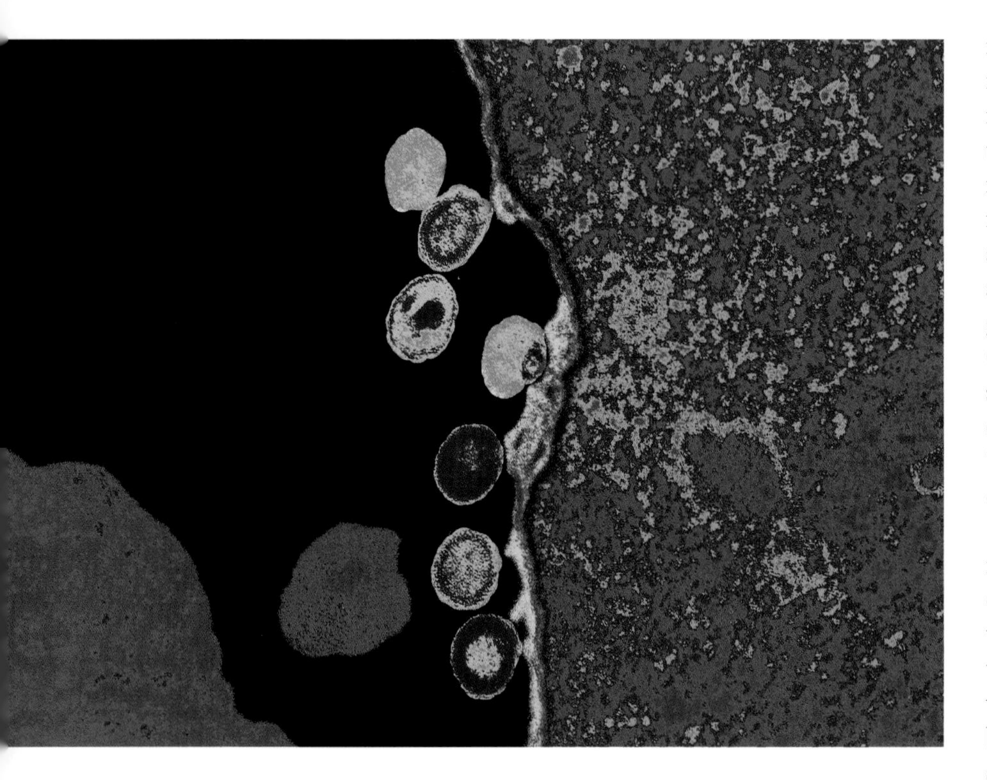

▲ *Virus particles of HIV leaving a cell. HIV may remain present in memory T cells, preventing drugs from eradicating the virus from the body.*

binding molecules is vast, but only those that are needed to deal with a clear threat are chosen, and the impact on the size of the genome is very modest.

Molecules of antibody, or immunoglobulin (Ig) are made in B cells. In an immature B cell one gene segment is selected at random for every two or three parts needed to build each of two peptide chains. The number of gene segments available for each part varies from 4 to about 65, so the total number of permutations in a given molecule is large, producing about 3 million potential antibody shapes. For a surface immunoglobulin (sIg), the completed protein will have two of each of the two peptide chains. This assembling of gene segments to arrive at a new compound gene is called somatic recombination. It is achieved by specific enzymes, and the sites at which they work are tightly controlled. The result will therefore always be recognizably a gene for an Ig.

Fine-tuning

Thus far, the mechanisms for producing Ig are very similar to those involved in TCR production. However, Ig molecules produced in B cells undergo a further two rounds of mutation that generate still greater specificity in their ability to bind antigen. The enzymes that control somatic recombination also introduce new nucleotides, that is, ones not found in the germ-line DNA (that which the animal inherits from its parents). These small alterations increase the diversity of compound genes produced by about 30 million times. Crucially, these nucleotide mutations affect sections of the gene that are eventually expressed as the CDRs, the highly variable "hands" of the Ig molecule, which consequently bind to specific antigen.

Thus the immune system has an overall repertoire of about 90 trillion (9×10^{13}) different antibody molecules. A young B cell will generate just one of these specificities, and will start by displaying it as sIg. Some of these young B cells will be deleted; others will never be needed and will therefore never give rise to any descendants.

Further changes in specificity will occur once an antibody is found to be useful. If a fit is made between sIg and an antigen experienced during infection, the B cell in question begins to divide, and its progeny will secrete more of the same antibody. The adaptability of the system does not end there.

IN FOCUS

Tolerance and clonal selection

Because the generation of antibody and TCR binding specificity is essentially random, the body will inevitably sometimes produce antibodies or TCRs that recognize its own cells or molecules. The adaptive immune system must "weed out" such agents before they have a chance to proliferate and begin a destructive attack on "self" tissues. In other words, the immune system must be continually kept in check to ensure that it remains tolerant of "self." Control is achieved by a process called clonal selection. Selection of T cells takes place in the thymus gland, which gets rid of any thymocyte whose receptor binds too well to "self" MHC, and also any which fails to bind to anything at all. Only cells that pass this test are allowed to go on maturing and dividing. In the bone marrow, a similar selection awaits B cells, based on their sIg's recognizing or not recognizing the "self" antigens in their environment. If the receptors on the sIg bind to "self," the B cell undergoes programmed cell death or apoptosis—effectively, it commits suicide.

In a process called somatic hypermutation, some activated B cells begin randomly introducing single nucleotide mutations into genes used to build the antibody in question. Many of these mutations will make the resultant antibody less of a match for the antigen. Cells with this hypermutation will fail to go on being stimulated by available antigen, and so production will cease. However some mutations may actually improve the fit the antibody can make with the antigen. B cells in which this happens will be maximally stimulated, and this new improved version will go into full production. The overall result of somatic hypermutation is that the antibody continues to be improved on and may become more effective as the immune response matures.

CLOSE-UP

Drifting and shifting

When the DNA in a virus mutates only slightly, the process is called antigenic drift. Large and dramatic mutations in the virus are known as antigenic shift. In the case of antigenic drift, the virus may still be genetically similar to the original. People may retain some antibody immunity against the virus, and a vaccine for the original form may provide some protection. When a virus undergoes antigenic shift, however, existing antibodies and vaccines are powerless to combat it because it is so different from its previous form. The influenza A virus frequently undergoes antigenic drift. Sometimes it also undergoes antigenic shift. Worldwide epidemics (pandemics) of influenza often follow an antigenic shift in the influenza A virus. In 1918–1919, at least 20 million people died in an influenza A pandemic. Influenza A pandemics occur every 10 to 40 years. The last one was in 1968.

Cells with long memories

There is another crucial element in the response of lymphocytes: memory. At stages in the production and fine-tuning of an immune response, a subset of the successful cells will become quiescent and remain, not as functioning antibody secreting cells, but as "watchdog" cells ready to be stimulated should the same antigen be encountered again in the future. As a result, the response to any further infections will be much better and faster than the first time, as the immune system has a head start in the form of mature B cells poised to produce the most effective antibody right away.

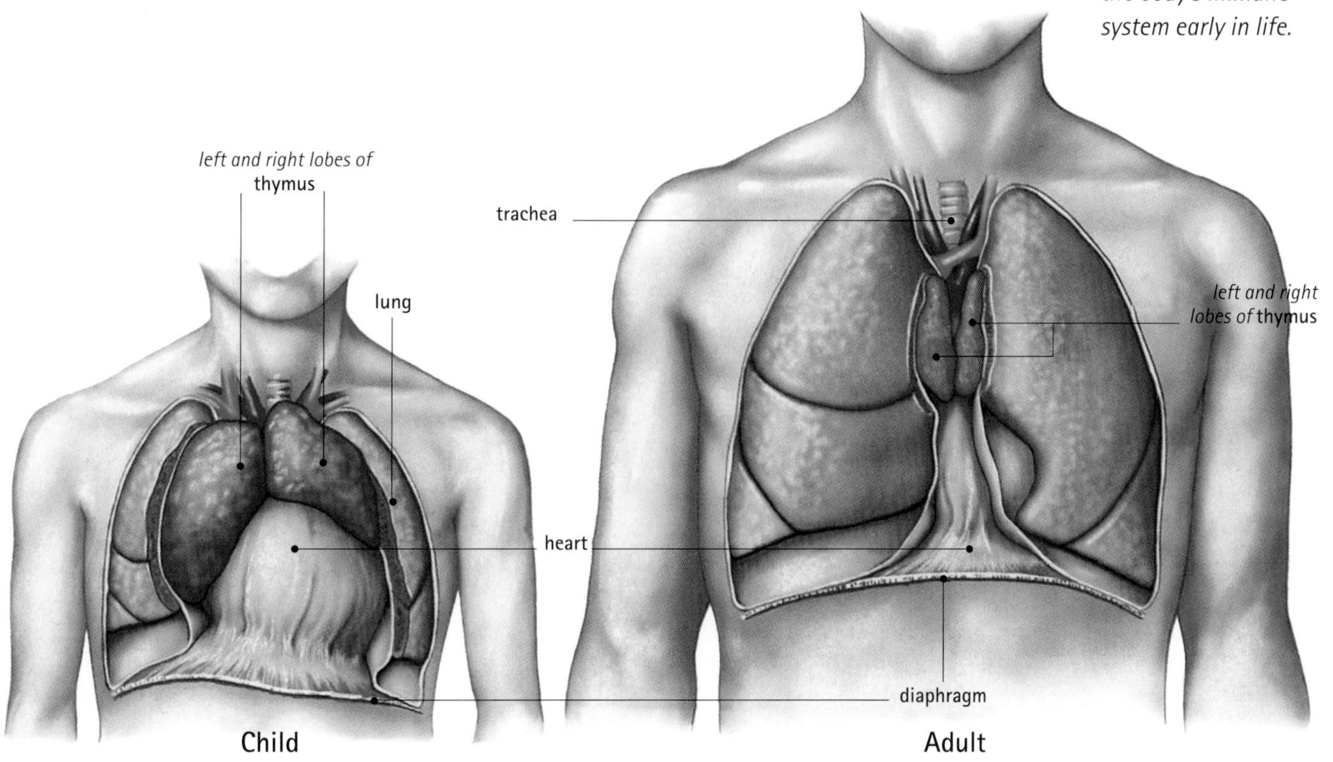

▼ SIZE AND LOCATION OF THYMUS
The thymus is much larger in a child than in an adult, reflecting the importance of this gland in establishing the body's immune system early in life.

Autoimmunity and genetic disorders

No biological system is perfect, and sometimes reactivity against self does occur. The unfortunate fact is that not all self-reactive B and T cells are excluded by clonal selection. The mechanisms that cause an immune system to attack the body it is supposed to defend are often far from clear, but a huge range of self material ranging from antibody molecules themselves to our own DNA can become targets of "autoantibodies." Similarly, T cells that damage self are frequently encountered. Some very serious and damaging diseases, including forms of diabetes and rheumatoid arthritis, are a result of autoimmunity. Surprisingly, the possession of autoantibody does not necessarily imply disease. Some autoimmune problems are temporary and seem to rectify themselves after a while. Thyroiditis, for example, which sometimes follows viral infection, is a short-lived disease, but the thyroid is also a frequent target of chronic (long-term) autoimmunity, which can be highly debilitating. It seems likely that at least a proportion of autoantibodies are a result of similarity between the specificities generated by particular infections and some self antigen. For example, a common complication of Chagas' disease, or American trypanosomiasis, is autoimmune damage to the nervous system, and streptococcal rheumatic fever appears to stimulate autoantibodies against heart tissues.

A genetic link?

Some autoimmune conditions appear to be associated with particular variants of MHC molecules. Because MHC molecules are coded for by germ-line DNA, such diseases can be inherited. Not everyone with a particular MHC variant gets the disease, but the association can be very high—as in the joint disease, ankylosing spondylitis. The root of the disease may be the way particular peptide binding is handled by individuals' MHC.

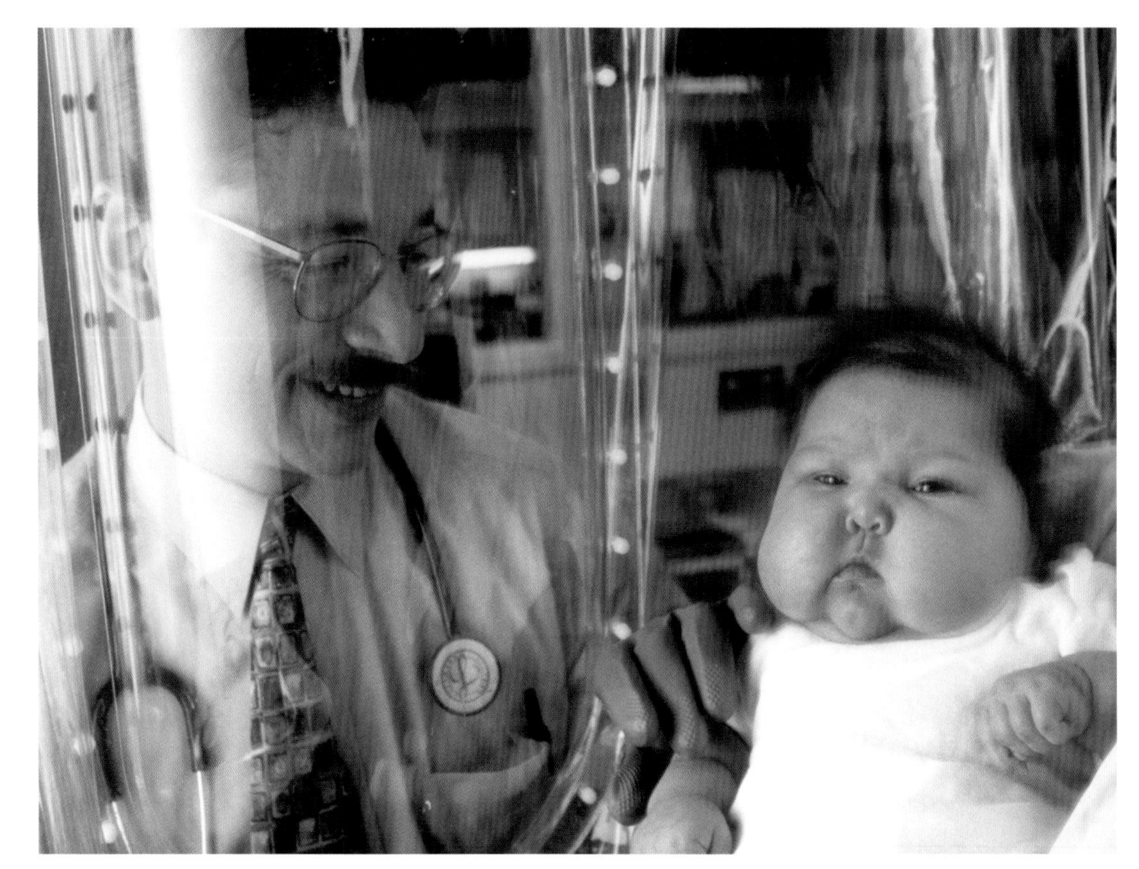

▶ *This five-month-old baby suffers from a condition called severe combined immunodeficiency (SCID). In this rare genetic disorder, babies fail to produce an enzyme vital to the functioning of the immune system.*

Unfortunately, some children are born lacking a component of the immune system. The first sign that an infant may have one of these rare congenital disorders will probably be the child's inability to fight a simple infection. Such disorders are called primary immunodeficiencies, and they include severe combined immunodeficiency (SCID), or "bubble boy disorder." Children with SCID have either no B cells and T cells or, if these cells are present, they have severe defects. Another congenital immune disorder is Wiskott-Aldrich syndrome (WAS). This syndrome is inherited through the X chromosome, so it affects only males, whose B and T cells can combat some infections but not others.

Apart from defects in lymphocytes, immunodeficiency can also be caused by numerous other failures: for example, the lack of a component of the complement cascade system, or a failure to produce the antibacterial toxins in killer granulocytes. People with genetic immune disorders may be treated with antibacterial and antiviral medications and are sometimes infused with purified antibody separated from donated blood. A number of attempts at gene therapy have been made: a modified virus is used to import a working copy of the defective gene into the patient's cells. This medical technology has proved unreliable, and sometimes dangerous, and it is now expected that stem cell transplantation will become the treatment of choice for many of these conditions.

GRAHAM MITCHELL/NATALIE GOLDSTEIN

FURTHER READING AND RESEARCH

Friedlander, Mark, and Terry M. Phillips. 1998. *The Immune System: Your Body's Disease-Fighting Army*. Lerner Publications: Minneapolis, MN.

Sompayrac, Lauren. 2003. *How the Immune System Works*. Blackwell: Malden, MA.

How your immune system works: http://health.howstuffworks.com/immune-system.htm

Immune system: www.niaid.nih.gov/final/immun/immun.htm

Immune system: http://uhaweb.hartford.edu/bugl/immune.htm

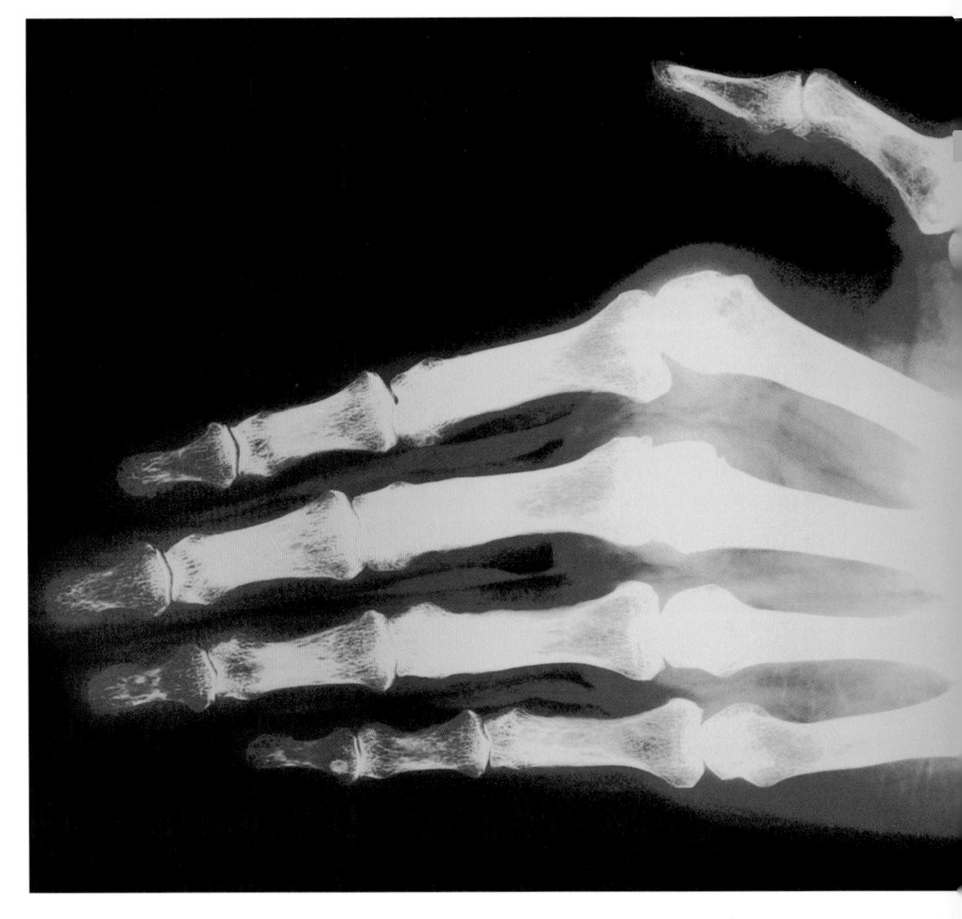

▲ *This heat-sensing image shows the warmth associated with inflamed, deformed, arthritic knuckles. Rheumatoid arthritis is an autoimmune disease in which the body's immune system attacks connective tissue, causing painful inflammation and stiffness in the joints.*

CLOSE-UP

One organ or more

There are two broad classes of autoimmune disorders: systemic and organ-specific. In systemic disorders, the immune system may attack multiple sites. The mechanism appears to involve a failure to clear immune complexes of antibody and self antigen from the circulation. One example of this is systemic lupus erythematosus, in which the immune system attacks joints, eyes, kidneys, and blood vessel walls. In organ-specific disorders, however, the immune system attacks only one specific type of organ. In the rare disorder called Addison's disease, for example, the immune system attacks only the adrenal cortex.

Jackson's chameleon

ORDER: Squamata SUBORDER: Sauria
FAMILY: Chamaeleonidae
GENUS AND SPECIES: *Chamaeleo jacksonii*

There are about 160 species of chameleons. More than half of these species are found only on the large island of Madagascar off the east coast of Africa. The remainder live in Africa, the Middle East, India, and Sri Lanka. Jackson's chameleon is found naturally in mountains in Kenya and Tanzania, but it has now established populations on some of the islands of Hawaii and in part of California.

Anatomy and taxonomy

Chameleons are reptiles. Other living reptiles include worm-lizards, snakes, turtles, crocodilians, and tuataras.

- **Animals** All animals are multicellular and feed off other organisms. They differ from other multicellular life-forms in their ability to move around (generally using muscles) and their rapid response to stimuli.

- **Chordates** Chordates are animals in which the long axis of the body is supported by a stiff rod called the notochord at some stage in the life cycle.

- **Vertebrates** The notochord of vertebrates develops into a backbone (also called spine or vertebral column) made up of units called vertebrae. The vertebrate muscular system that moves the head, trunk, and limbs consists primarily of muscles that are bilaterally symmetrical around the skeletal axis—in other words, the muscles on one side of the backbone are the mirror image of those on the other side.

- **Reptiles** Reptiles are vertebrates with thick, horny, waterproof skin that is usually divided into plates called scales. Most reptiles have four legs, and most lay eggs with a waterproof shell. Reptiles are not endothermic (warm-blooded) like birds and mammals, but they can exert some control over their body temperature by moving into hot places when they need to warm up and into cooler ones when they need to lose heat.

Animals
KINGDOM Animalia

Chordates
PHYLUM Chordata

Vertebrates
SUBPHYLUM Vertebrata (or Craniata)

Reptiles
CLASS Reptilia

Turtles
ORDER Chelonia

Squamates
ORDER Squamata

Crocodilians
ORDER Crocodilia

Snakes
SUBORDER Serpentes

Lizards
SUBORDER Sauria

Worm-lizards
SUBORDER Amphisloaemia

Chameleons
FAMILY Chamaeleonidae

Other chameleons
SUBFAMILY Chamaeleoninae

Jackson's chameleon
GENUS AND SPECIES *Chamaeleo jacksonii*

Dwarf chameleons
SUBFAMILY Brookesiinae

◀ *The family Chamaeleonidae (chameleons) is one of 17 families in the reptile order Squamata, the snakes and lizards.*

▲ *Most chameleons live in tropical regions of Africa and Asia. Remarkable for their ability to change color quickly, chameleons have a prehensile tail, which they use as a balancing aid.*

• **Lizards** Most lizards are long and thin with four legs. Some species, however, do not have legs and so look superficially like snakes. Externally, the main differences between lizards and snakes are that lizards have eyelids (snakes do not and so cannot close their eyes) and an ear opening (in snakes the ear is entirely internal), and there are several rows of scales on the underside of the body (snakes have only one row). Since they are similar in many respects, lizards and snakes are placed in the same order, Squamata. Lizards are assigned to the suborder Sauria and snakes to the suborder Serpentes.

• **Chameleons** The sides of a chameleon's body are relatively flat, and in many species the tail is prehensile—that is, it can be coiled around twigs or branches to help support the lizard. The pupils of the eyes are at the top of conical "turrets," which can move independently of each other; one eye may look in one direction while the other looks the other way. The fingers and toes on each of a chameleon's four legs are fused to form two grasping pads. Most chameleons live in trees, and the grasping pads help them hold branches very firmly. A chameleon's very long tongue has a sticky tip and can be projected very rapidly from the front of the mouth to capture insects and other invertebrates. Chameleons are able to change their color and color pattern more completely and quickly than any other vertebrate.

FEATURED SYSTEMS

EXTERNAL ANATOMY Chameleons have a flat-sided body. Their eyes are borne on scaly "turrets," which can move independently of one another. Chameleons' feet are arranged so that two fused digits oppose three fused digits, enabling them to exert a powerful grip. Their tail is usually coiled. *See pages 626–627.*

SKELETAL SYSTEM The back of the skull is elongated to support a "casque." Male Jackson's chameleons have three "horns" at the front of the head, which are supported by bony extensions of the skull. *See pages 628–629.*

MUSCULAR SYSTEM The muscles of chameleons contract more slowly than those of other lizards, so movements are slow. *See page 630.*

NERVOUS SYSTEM Chameleons are able to change color in a dramatic way. This is caused by changes to pigment-containing cells in the skin, which are controlled by the nervous system. *See page 631.*

CIRCULATORY AND RESPIRATORY SYSTEMS The lungs have hollow projections that help chameleons inflate their body and intimidate rivals or predators. *See page 632.*

DIGESTIVE AND EXCRETORY SYSTEMS The tongue can be extended to more than the length of the body and is used for capturing prey. *See page 633.*

REPRODUCTIVE SYSTEM Female Jackson's chameleons retain their eggs in the oviducts until the embryos are ready to hatch. *See pages 634–635.*

External anatomy

The **eyes** *are set in swiveling turrets. This arrangement gives chameleons all-around vision.*

Male Jackson's chameleons use their **horns** *as weapons in territorial clashes. The horns may also help break up the chameleons' outline, providing camouflage.*

The **tongue** *can be propelled very quickly from the mouth. The tip of the tongue is coated with a viscous liquid that ensnares insect prey. When fully extended, the tongue is longer than the chameleon.*

CONNECTIONS

COMPARE the scales on a chameleon with those on a ***CROCODILE.***

COMPARE the opposable digits of a chameleon with those of ***MANDRILL.***

▶ **Male Jackson's chameleon**

Male Jackson's chameleons have three large horns, one from the snout and two above the eyes. Females usually have just one tiny horn on the snout.

Chameleons are very unusual lizards. Most chameleons live in trees, bushes, or other dense vegetation where their flat-sided body helps to conceal them among the leaves. The tail, however, is not flat. It is usually coiled up and can be curled around twigs or branches to help support and stabilize the lizard. The eyes are in the middle of scaly conical "turrets." When chameleons change the direction in which their eyes are pointing, the whole "turret" moves. Each eye can move independently, so that one eye can be looking in one direction while the other is looking in another. Chameleons, unlike most other lizards, have no external ear opening. The arrangement of the digits (toes) enables chameleons to grasp small branches. Two fused digits oppose three fused digits, so that the lizard can exert a powerful grip.

CLOSE-UP

Fearsome horns

Male Jackson's chameleons have three horns projecting forward from the head. One horn extends from the snout, immediately above the upper jaw. The other two, which are shorter, lie approximately side by side between the eyes. The horns are supported by bone and covered in scales arranged in neat rings. The horns are usually reddish brown and, unlike other scales on the body, are not capable of changing color. Female Jackson's chameleons have only a single very small horn. The top of the chameleons' head has a helmetlike extension, called a casque. Like the horns, the casque is supported by bone. The scales, however, are irregular. Male Jackson's chameleons maintain territories, and the horns are used as weapons when disputes about territories result in fights. The casques, however, are probably bluff. They make the head of the lizard seem larger, and thus more fearsome, than it actually is.

The skin of chameleons is covered in small horny scales. In many lizards the scales run in neat rows, but in chameleons they are mostly irregular. They do not overlap one another at any point, whereas many lizards have at least some scales that do. The scales on the head are slightly larger than those on the body. Along the middle of the back is a crest of pointed scales, and these also run along the top of the tail. They are smaller in Jackson's chameleon than in many other species. The scales lining the upper and lower jaw are slightly larger than the remaining scales on the head, and they form a neat row. Some of the scales on the body are larger than the rest and are called tubercles. These have an irregular distribution, unlike the tubercles on some other lizards and on crocodiles, which are in neat rows.

The **skin** *contains pigment cells that are controlled by the animal's nervous system. These cells can quickly alter the color, hue, and pattern of the skin. Chameleons regularly shed their old skin.*

The **tail** *is cylindrical and prehensile. It is generally coiled tightly when not in use, but it can act as a fifth foot.*

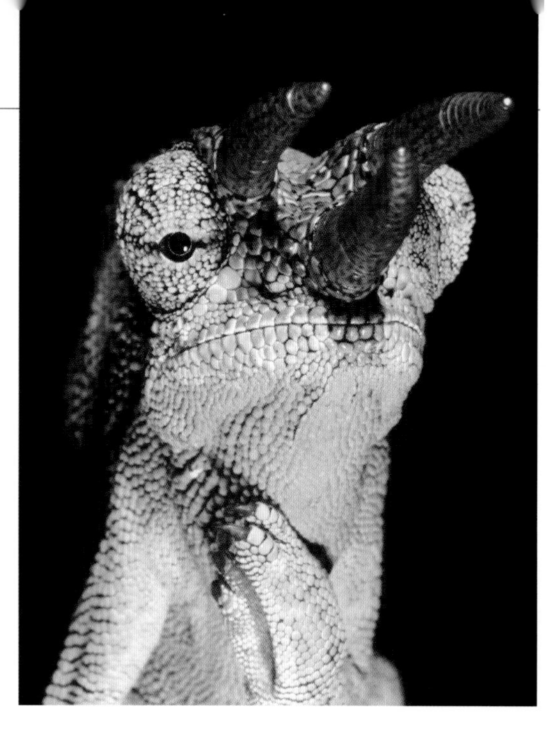

▶ *Like the horns of a male Jackson's chameleon, those of Johnston's chameleon are usually slightly curved and are covered with rings of scales. The scales on the horns are not capable of changing color.*

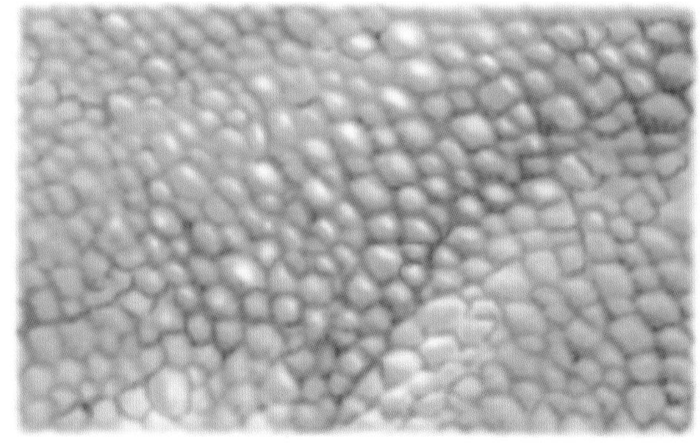

▶ SCALES
Chameleon
A chameleon's scales are small and do not overlap each other. Although they are irregular in size, they are not armored.

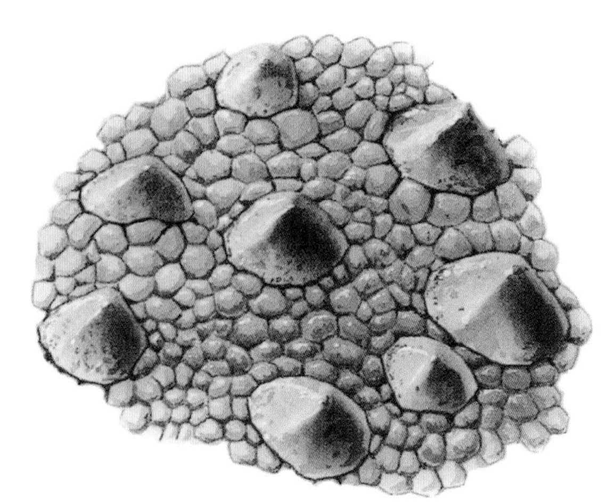

◀ SCALES
Iguana
Iguanas, like chameleons, are lizards, but their scales are different. Iguana skin is armored and covered with scales of very different sizes.

Skeletal system

▶ *The skeleton is similar to that of other lizards, but two features are very distinct: the long horns of an adult male and the large orbits to accommodate the eyes.*

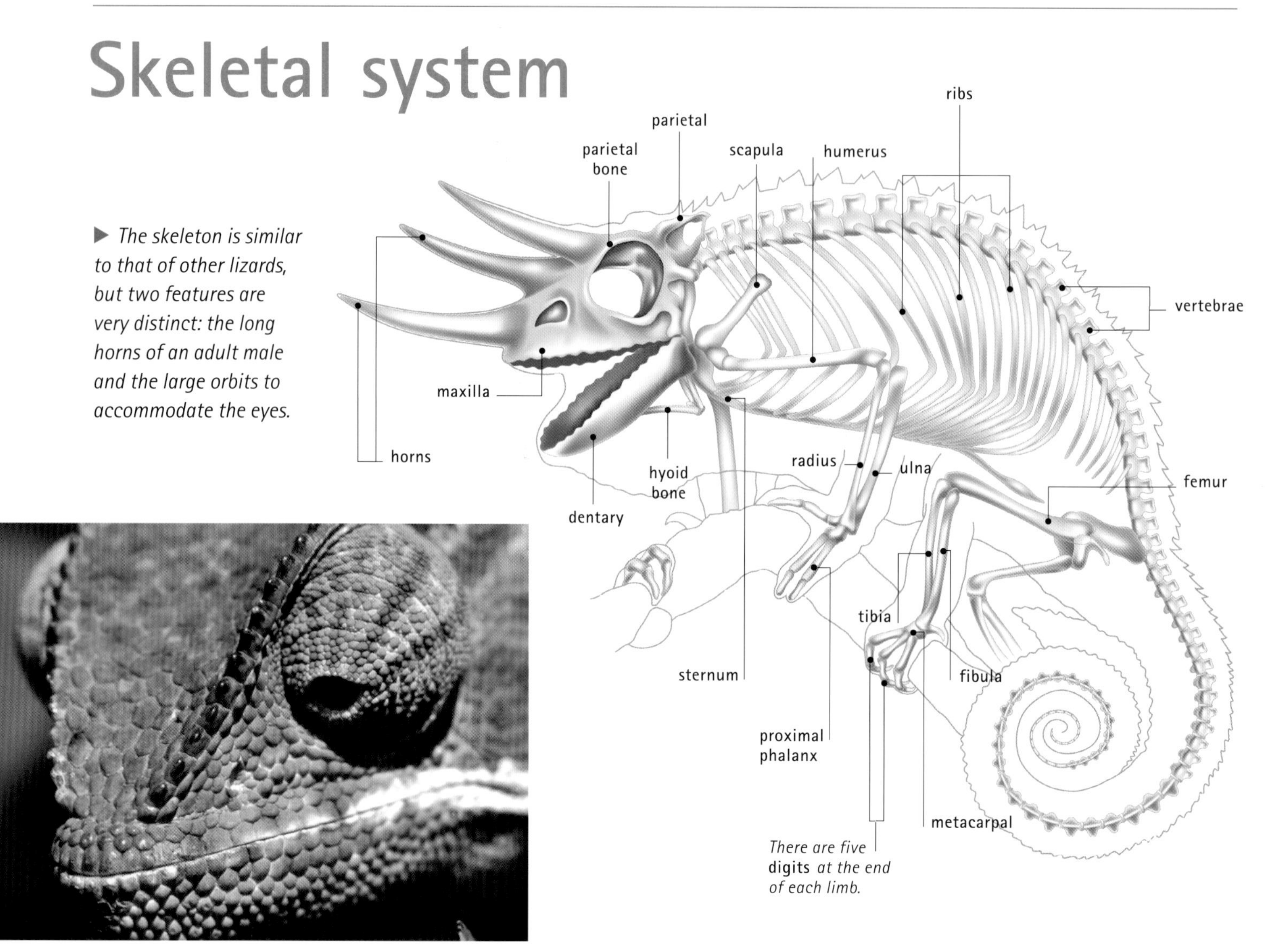

▲ *Chameleons' eyes look sideways. Chameleons cannot judge distances accurately using binocular vision like many other vertebrates. Experiments have shown that when a chameleon focuses on an object using muscles in its eyes, the extent to which the muscles contract is used to judge distances. Although chameleons are not the only vertebrates that can do this, the ability is uncommon.*

Lizards, alligators, crocodiles, and tuataras are all reptiles with similar skeletons, although experts are able to recognize detailed differences between the groups. In contrast, snakes have an elongated skeleton with very large numbers of vertebrae and ribs, and they have no limbs. Turtles have fused ribs that support their shell.

Compared with the skull of a "typical" lizard, a chameleon's skull has very large spaces, or orbits, for the eyes. The large orbits enable the skull to accommodate the unique "turrets" surrounding the front of each eyeball. The parietal bone at the back of the skull and the two squamosal bones are elongated upward and backward. It is difficult to separate the individual bones in the skull of a chameleon because many of them have become fused with one another. The parietal and squamosal bones together form a tripod that supports the casque, a hump on the back of the head. The casque probably intimidates other chameleons and perhaps also potential predators.

Jackson's chameleons have three horns at the front of the head. These are especially large in adult males. The horns are supported by extensions of bones at the front of the skull. The larger, central horn on the snout is supported by an extension of the premaxilla and the two premaxillary bones. The other two horns are supported by extensions of the prefrontal bones.

Tiny teeth

Chameleons have rows of very small teeth. Jackson's chameleons have 18 teeth on each side of the upper jaw fixed to the maxilla and one tooth on each side fixed to the premaxilla. They have 18 teeth on each side on the lower jaw, which are fixed to the dentary bones.

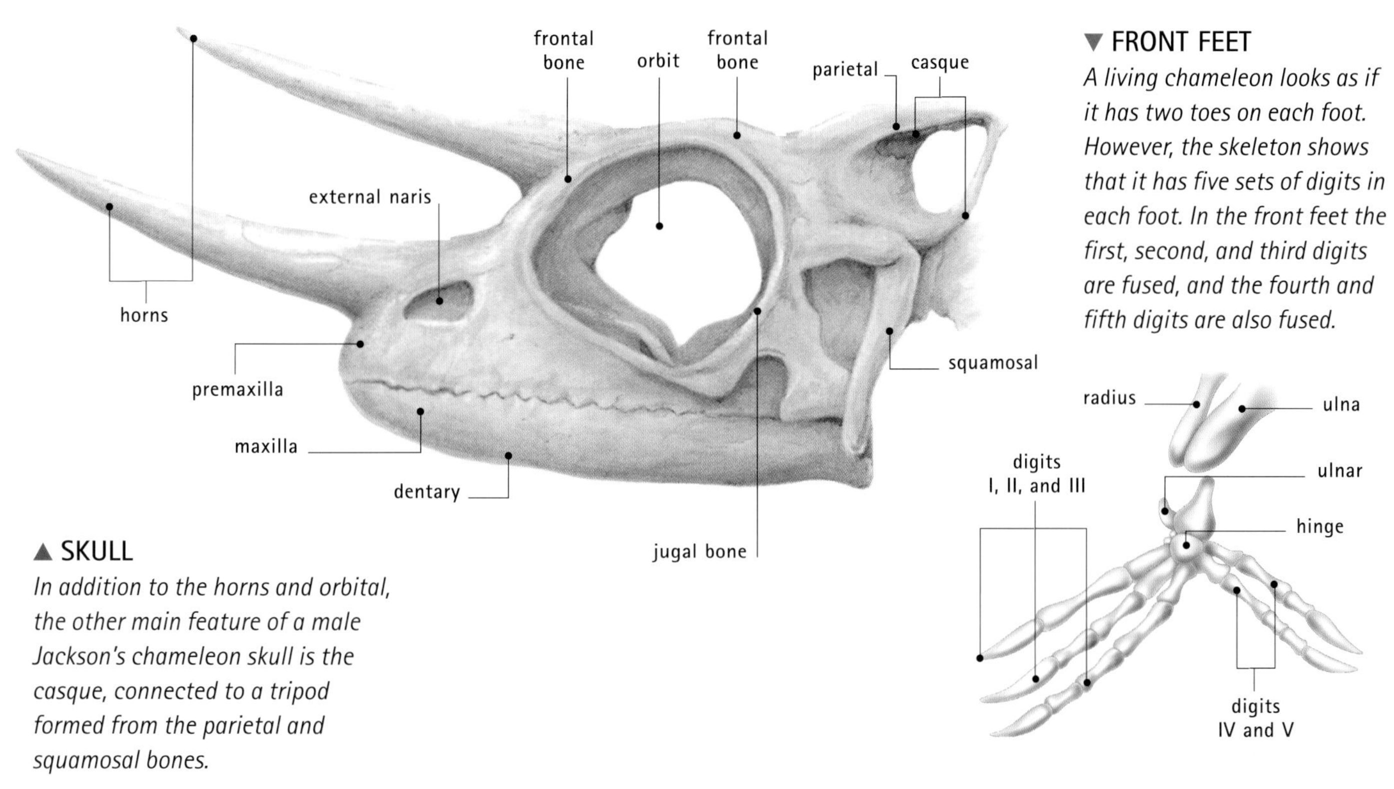

▲ **SKULL**
In addition to the horns and orbital, the other main feature of a male Jackson's chameleon skull is the casque, connected to a tripod formed from the parietal and squamosal bones.

▼ **FRONT FEET**
A living chameleon looks as if it has two toes on each foot. However, the skeleton shows that it has five sets of digits in each foot. In the front feet the first, second, and third digits are fused, and the fourth and fifth digits are also fused.

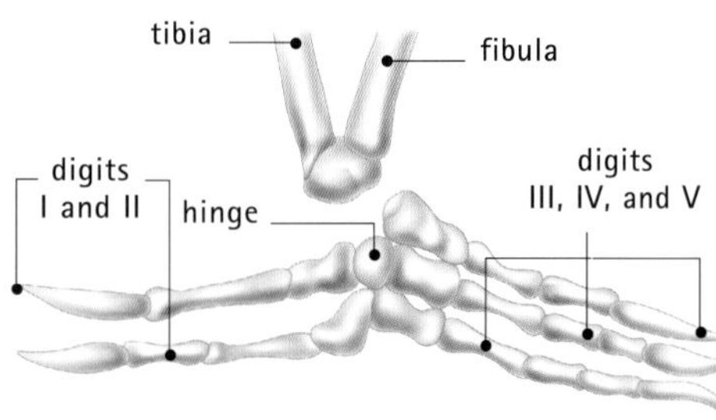

◀ **HIND FEET**
The first and second and the third, fourth, and fifth digits are fused. The digits rotate around a hinge. The individual bones of each digit are called phalanges.

Grasping fingers and toes

The bones of the legs, hands, and feet of amphibians, reptiles, and mammals have the same basic plan. Birds also have the same plan, but in birds the front legs and feet form wings. Scientists call the plan the pentadactyl limb because there are usually five fingers or toes on each leg. Chameleons' limbs fit the pentadactyl plan. The fingers and toes are made up of a number of individual bones called phalanges. In the human hand, the first finger (the thumb) opposes the remaining four. When we grasp an object, we squeeze it between the thumb and the remaining fingers—usually the second. On a chameleon, the first three digits of the front legs are fused, and they oppose the fourth and fifth digits, which are also fused. A chameleon looks as though it has two digits, but the bones show that they are derived from five.

On the hind legs, the first and second digits are fused; they oppose the fused third, fourth, and fifth digits. Both arrangements make the grip very strong. This strength is helped by the arrangement of bones in the wrist. The "hinge" for each hand or foot is a single wrist bone. In the front leg this is the carpal bone; in the back leg it is the tarsal bone.

COMPARATIVE ANATOMY

Tails that can drop off

Most lizards are able to lose their tail, but chameleons do not have this ability, called autotomy. Autotomy usually occurs if the tail is grabbed by a predator: it drops off, the lizard survives, and a new tail grows from the stump where the previous one was severed. It might be expected that the breakage occurs when two vertebrae pull part, but the break does not happen in that way. Each vertebra grows with a small crack in it called a fracture plane. The separation of the bone at a crack enables the tail to drop off very easily. Although the lizard grows a new tail, it is never as long or brightly colored as the original. Therefore, a lizard with a regrown tail is easy to recognize. Scientists are very interested in the ability of lizards to grow a new tail because they are the most advanced vertebrates able to regenerate a whole organ. Chameleons cannot regrow their tail, but the tail is more important for them than it is for other lizards. Chameleons are able to coil the tail around twigs and branches to give extra support; a grasping tail is called prehensile.

Muscular system

CONNECTIONS

COMPARE the prehensile tail of a chameleon with that of a ***SEA HORSE.***

COMPARE the tongue muscles of a chameleon with those a ***GIANT ANTEATER.***

The most unusual part of the muscular system of chameleons is the arrangement of muscles along the underside of the tail. These muscles enable the tail to coil up and twist around twigs or branches, giving extra support. The actual arrangement of the muscles is very complex. Chameleons move very slowly, and sometimes two legs are raised at the same time. Without the support offered by the tail, this method of locomotion would make the animal unstable: it could easily fall.

Muscles aid camouflage

Chameleons do not need to move fast because they stalk their prey until they are near enough to shoot out their long sticky tongue and pull it into the mouth. Slow movement also helps chameleons blend with the surrounding leaves, and some chameleons move irregularly and with lots of pauses; their progress is a series of short lunges interspersed with periods when they are immobile. Many chameleons even rock backward and forward. Their movements mimic those of leaves in a breeze. Combined with their thin body shape and their color, this behavioral camouflage makes them harder for predators to see.

IN FOCUS

Muscle cells

The study of how cells are constructed and arranged to form organs is called histology. Muscle cells have a complicated histology. They are long and thin, and lie together in groups called fibers. Most of the muscles of vertebrates have a banded appearance when viewed with a microscope and are called striated muscle cells. There are two basic types of striated muscle fibers. Some can contract very rapidly; others contract more slowly but exert more power and have more stamina. Humans vary in the proportions of fast and slow fibers in their muscles. Olympic sprinters have a high proportion of fast fibers, and marathon runners have a high proportion of slow fibers. Most of the muscles of chameleons also have a very high proportion of slow fibers because they are built for stamina, not for speed.

▼ **TONGUE MUSCLES**

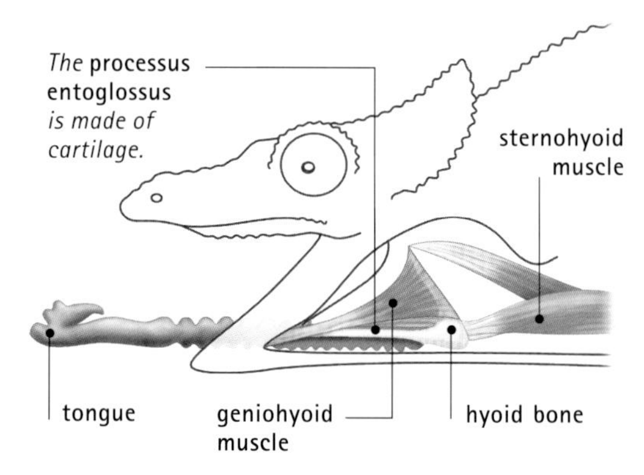

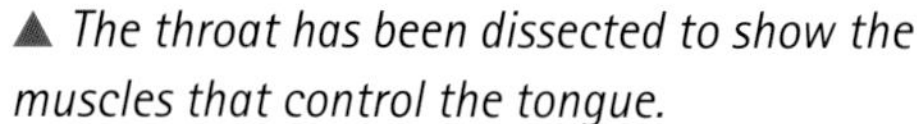

▲ *The throat has been dissected to show the muscles that control the tongue.*

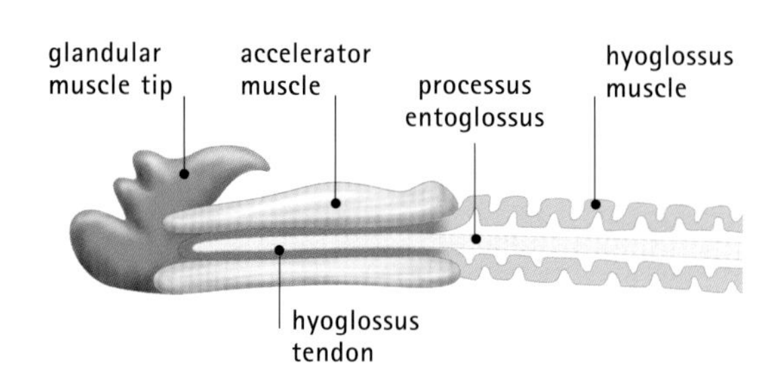

▲ *This section shows the arrangement of muscles when the tongue is retracted. The hollow hyoglossus tendon surrounds the slippery processus entoglossus.*

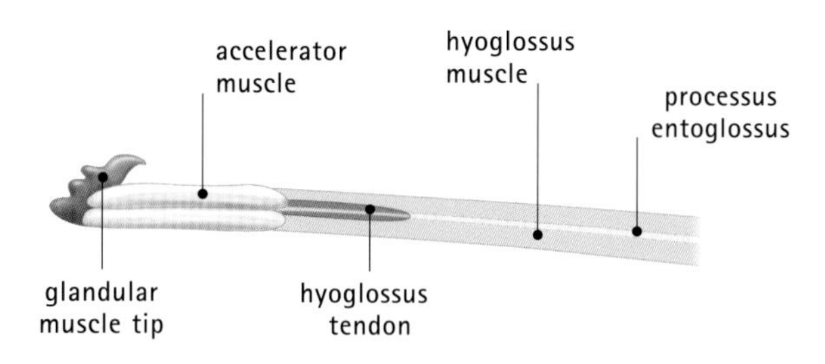

▲ *This section shows the tongue projected and with the pleatings of the hyoglossus stretched out. The hyoglossus tendon is pulled off the processus entoglossus and turned inside out.*

Nervous system

The most striking feature of chameleons is their amazing ability to change the color and pattern of their skin. They can do this more completely and more rapidly than any other vertebrate. The most usual color of a Jackson's chameleon is a shade of green, but there are usually patches of darker green, brown, reddish brown, black, and yellow dotted around the body. Jackson's chameleons are usually paler in overall color at night than during the day. Changes in color and color pattern can occur as a result of changes in the background environment or temperature, or as a result of social interactions, especially when males fight or during courtship.

How chameleons change color

The skin of lizards has two main layers: an outer epidermis where the cells are continually dividing so that they are replaced at intervals; and an inner layer called the dermis, which is not shed at intervals. Some of the cells in the dermis contain pigment granules. Those cells containing black or dark brown pigment are the most common. They are called melanophores. These cells can change their shape, and can extrude long projections into the surrounding tissue. When melanophores are expanded, the pigment granules are dispersed over a wide area, and the skin will appear dark. When they contract, the pigment is condensed, and the skin appears pale.

This explains changes in shade. Changes in color can occur because there are other sorts of pigment-containing cells, too. The most important ones contain red pigment or yellow pigment. Others contain crystals that reflect light. Cells that contain colored pigments are collectively called chromatophores. These cells are able to move within the dermis. Changes in color are produced in two ways. First, if the melanophores expand, the chromatophores nearer to the skin surface are obscured. Second, if cells containing, for example, yellow pigment move within the dermis and come to lie near the outside, they act as a filter on light reflected from the crystal-containing cells, and the skin appears green.

This is a very simplified account of an extremely complicated mechanism. This mechanism for color change is not unique to chameleons. Almost all fish, amphibians, and reptiles have the ability to alter their color. The feature that gives chameleons such amazing control over the process is that the movement of melanophores and chromatophores, and their expansion and contraction, is mostly controlled by the nervous system. In most other vertebrates these processes are controlled mostly by hormones—indeed, there is a hormone produced from the pituitary gland at the base of the brain in vertebrates whose main function is to control melanophores, and so—not surprisingly—it is called melanophore-stimulating hormone (MSH).

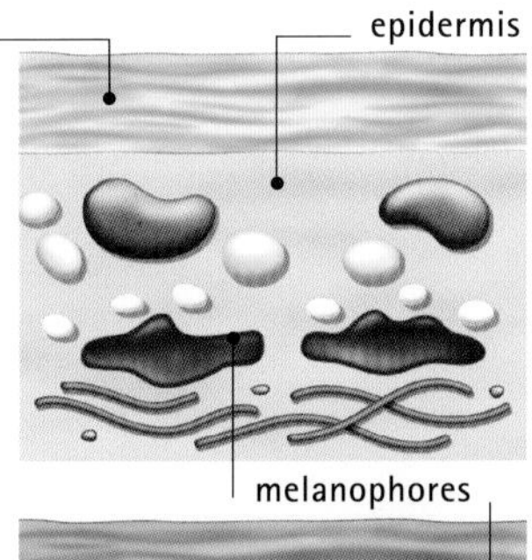

▶ *Red and yellow chromatophores dominate near the surface of the dermis; the lizard appears orange.*

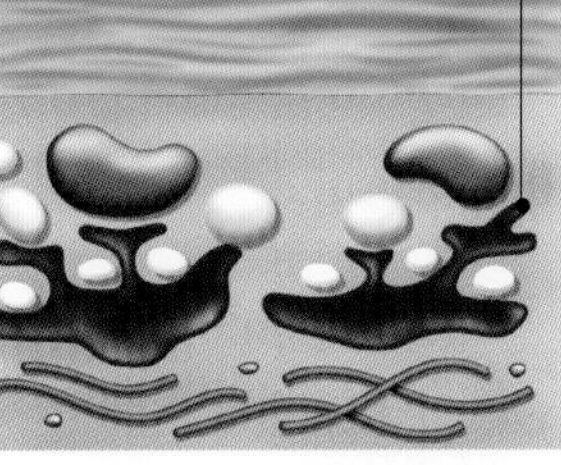

▶ *Black and dark brown cells called melanophores move toward the bottom of the epidermis, causing the chameleon to appear terra-cotta.*

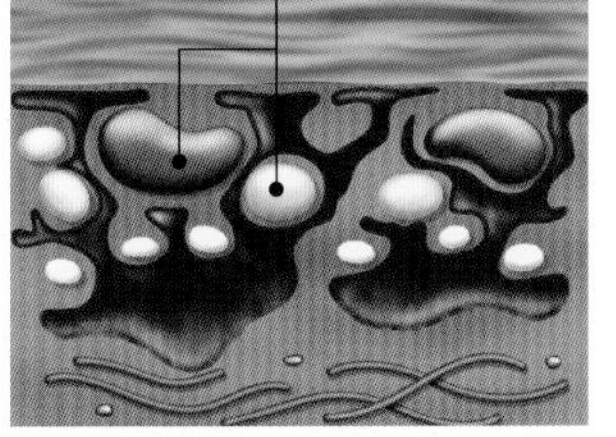

▶ *The melanophores spread out under the epidermis, and the lizard darkens to chocolate brown.*

◀ **CHROMATOPHORES**
Chameleon, gecko, or agama
The diagrams show the process whereby lizards are able to change the color of their skin.

CONNECTIONS

COMPARE the color-changing ability of a Jackson's chameleon with that of an ***OCTOPUS***.

COMPARE the camouflage of a chameleon with that of a ***BULLFROG***.

Circulatory and respiratory systems

CONNECTIONS

COMPARE the breathing system of a chameleon with that of a ***HUMAN.***

COMPARE the heart of a chameleon with that of a mammal such as a ***GRAY WHALE.*** Mammals have a four-chamber heart, but lizard hearts have only three chambers.

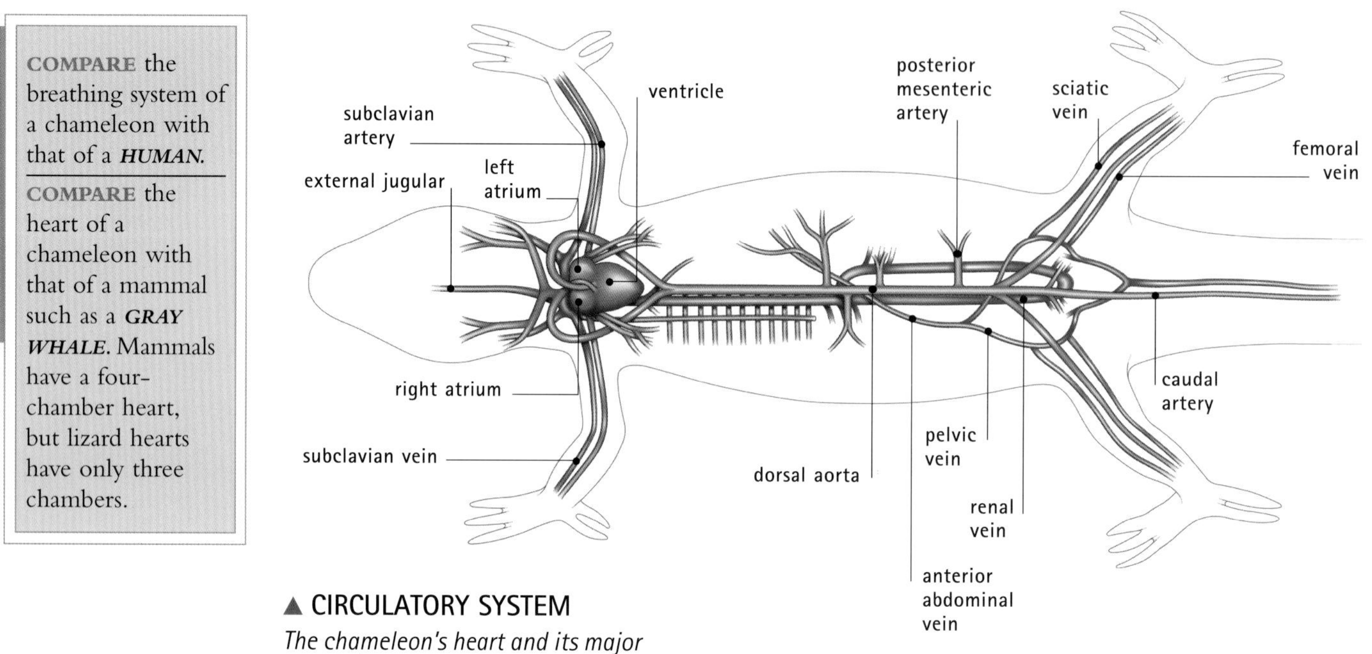

▲ CIRCULATORY SYSTEM
The chameleon's heart and its major arteries and veins, viewed from below.

Although chameleons possess many unusual anatomical features, their circulatory and respiratory systems are almost identical to those in all lizards. The heart has three chambers. Two of the chambers, the left atrium and the right atrium, lead into the third chamber, the ventricle. Oxygen-rich blood from the lungs enters the left atrium and is pumped into the ventricle. The arteries that lead to the tissues of the body (the systemic arteries) are on the left side of the ventricle. Therefore, most of the oxygen-rich blood from the left atrium passes into the systemic arteries, then into the carotid arteries, and is carried to the tissues of the body. Having given up its oxygen in the tissues, blood returns to the right atrium of the heart through the venous system and is pumped again into the ventricle. The deoxygenated blood enters the right side of the ventricle, and most of it passes to the pulmonary arteries situated on the right side of the ventricle and so again to the lung.

IN FOCUS

Puffed up like a balloon

To intimidate an aggressor or a potential predator, chameleons inflate their lungs to the maximum extent, and so their body gets bigger. To increase the effect, the lungs of chameleons have a number of projections that extend into the body cavity. The projections are not used for breathing but to enable more of the body to be filled with air. Some lizards use this kind of body inflation for protection. They go into a crevice in the rocks and inflate their body so they become wedged in.

Endurance capacity

Unlike mammals, lizards do not have a diaphragm separating the lungs from the remainder of the abdominal cavity. In lizards, pumping air into the lungs and expelling it after some of the oxygen has been absorbed by the hemoglobin in the red blood cells is achieved by complicated movements of muscles of the ribs, in the neck, and on the floor of the mouth cavity. This is not a very efficient system and is one of the reasons why lizards cannot run fast for very long. Their respiratory system cannot deliver oxygen to the tissues as fast as it is being used.

Digestive and excretory systems

◀ *A chameleon projects its tongue to catch a passing insect. From the time the chameleon shoots out its tongue to the time the prey is safely trapped in the reptile's mouth takes less than a second. The prey will be passed from the mouth to the stomach and then to the intestines, where digestion occurs.*

Vertebrates that feed mostly on plants have relatively longer and more complicated intestines than vertebrates that feed on meat. This is because plant cells are much harder to digest than animal cells. Chameleons feed exclusively on insects and other small invertebrates, and their intestinal system is therefore a simple one. Food is moved from the mouth along the esophagus to the stomach. The wall of the stomach is extremely elastic, and so can accommodate relatively large items of food. Chameleons sometimes swallow hawkmoths so large that it seems impossible that they can stuff the whole insect in their mouth. Food passes from the stomach to the intestines, where most digestion occurs; then waste is moved from the large intestine to the cloaca, the combined digestive and reproductive opening, from where it is expelled.

Projectile tongue

Chameleons' projectile tongue is unique among vertebrates. It can be shot out at high speed to catch insect prey at a distance of about one body length. As described earlier (in the muscular system section) the tongue is sticky. Its tip has glands that produce a sticky substance called mucus. The insect adheres to this and is then drawn into the mouth and swallowed. The tongue is tubular and contains a rod made of a strong but flexible material called cartilage. The rod, or processus entoglossus, runs down the center of the tongue. The front part of the tongue has a tubular muscle with circular fibers running around the outside. This is called the accelerator muscle. The remainder is made up of a muscle with longitudinal fibers called the hyoglossus. When the tongue is not in use, this muscle is folded like a concertina. When the accelerator muscle contracts, it squeezes the processus entoglossus, which shoots forward at high speed, carrying the remainder of the tongue with it. Where the thickened end of the tongue fits over the processus entoglossus, it is attached by a tubular tendon, which runs from the front of the processus to the back of the club-shaped part of the tongue. This tendon is turned inside out, increasing yet farther the distance the tongue can be projected. The tongue is withdrawn by contraction of the hyoglossus muscle.

Reproductive system

CONNECTIONS

COMPARE the structure of a chameleon egg with that of an ***OSTRICH.***

COMPARE the chameleon's placenta, formed from the chorion and allantois, with the placenta of a placental mammal such as a ***HUMAN.***

Jackson's chameleons breed once a year. Females can begin breeding when they are just over one year of age, but males do not begin to produce sperm until they are nearly two. The female retains the eggs in the oviducts until they are ready to hatch. The young are born some time between mid-January and mid-March, which is the wet season in the areas of Kenya and Tanzania where the species lives. Many of the insects on which they feed also breed in the wet season, so birth coincides with the maximum abundance of food. When the eggs are ready to hatch, the female chameleon descends to the ground, and the young chameleons, which are 1.6 to 1.8 inches (4 to 4.5 cm) long, are born through the vent. Each newborn chameleon is surrounded by a transparent membrane, but the baby ruptures this and breaks free almost immediately. A female produces an average of about 20 young.

The two ovaries of female chameleons lie at the back of the abdominal cavity. Ovaries produce ova, which are the cells that will eventually multiply after fertilization to form an embryo. Each ovum develops in a space within the ovary called a follicle. Eventually the ova burst out of their follicle, and lie in the tubes that connect the ovaries to the cloaca at the back of the alimentary canal. The tubes are called oviducts. Development of each single-celled ovum into an embryo takes place within the oviducts in those species, such as Jackson's chameleon, that produce live young.

Fertilization of the ova by sperm produced in the two testes of the male chameleon takes place in the oviducts. All lizards and snakes have two penises, or hemipenises, one on either side of the cloaca. Only one, however, is inserted into the cloaca of a female during mating. Sperm develop within the testes in cells in the walls of small tubules. The sperm pass to a coiled tube that runs across the outside of each testis. The coiled tube is called the epididymis and is where spermatozoa are stored until they pass into one of the hemipenes during mating. Each epididymis is connected to the hemipenis by a tube called the vas deferens.

▼ During mating, the male inserts one of his two hemipenises into the female's cloaca. Sperm produced in the male's testes is ejaculated into the female's cloaca and fertilizes the eggs in the oviducts.

Seasonal or continuous breeding?

Most species of reptiles have a seasonal reproductive cycle. Reproductive cycles are controlled mainly by changes in day length, temperature, and humidity. These are registered in a part of the midbrain called the hypothalamus, which controls secretions of hormones from the pituitary gland. The reproductive hormones then act on the gonads (the testes and ovaries) to control the timing of the breeding cycle.

Some reptiles, however, produce ova and sperm throughout the year. All those that have a continuous breeding cycle live in the tropics. They can mate, and the females can produce clutches of eggs or living young, at any season. Hoehnel's chameleon, which lives in the same areas of Kenya and Tanzania as Jackson's chameleon, is a continuous breeder. Any individual female Hoehnel's chameleon produces two clutches of young a year, but the timing is not synchronized among females, so young are being born all the time. After mating, a female Hoehnel's chameleon is able

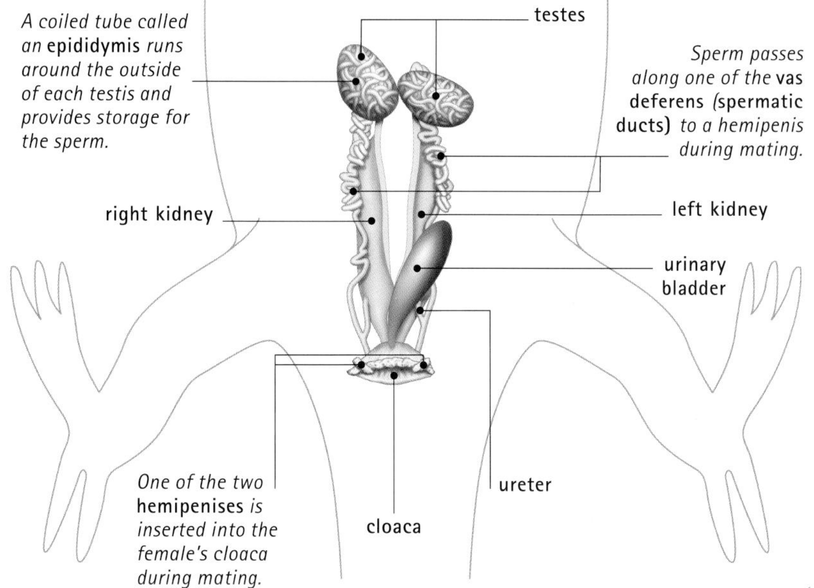

◀ **MALE UROGENITAL SYSTEM**
This is an underside view, showing the reproductive organs and the main features of the urinary system.

▼ **FEMALE UROGENITAL SYSTEM**
This underside view show the component parts of the reproductive and urinary systems, which share a common opening—the cloaca.

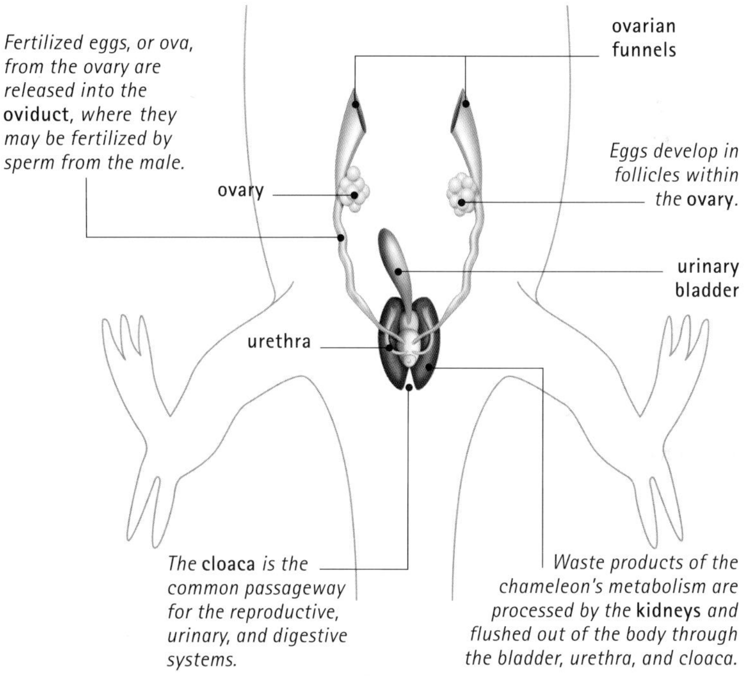

to store sperm within the oviducts. The sperm can remain viable for at least a year, and sometimes longer. When the female is ready to reproduce, the sperm are used to fertilize the eggs. Thus it is not necessary for a female Hoehnel's chameleon to mate immediately before the fertilization of her eggs.

ROGER AVERY

FURTHER READING AND RESEARCH

Harris, T. (ed.). 2003. *Reptiles and Amphibians.* Marshall Cavendish: Tarrytown, NY.

Mattison, C. 1989. *Lizards of the World.* Facts on File: New York.

Smith, H. M. 1995. *Handbook of Lizards of the United States and Canada.* Cornell University Press: Ithaca, NY.

CLOSE-UP

Egg membranes

The females of most species of reptiles lay eggs. All reptile eggs are laid on land, and they face a number of problems. The developing embryos need protection, not only from physical damage but also from drying up. They need to be nourished and to be able to breathe. Another problem is that they need to be protected from the products of the break-down of proteins during normal metabolism. These products always contain nitrogen and are toxic.

The developing embryo within a reptile egg comes to be attached to four membranes. Two of them envelop it. The spaces between these two membranes and between the inner membrane and the embryo itself are filled with fluid. These two membranes are called the amnion and the chorion. The fluid cushions the embryo from possible damage. The third membrane forms a sac, which is filled with yolk. This is the food for the developing embryo. By the time that the embryo hatches, very little yolk is left. The fourth membrane also forms a sac. It is called the allantois, and it acts as a repository for nitrogenous excretory waste, so that this waste does not come into contact with the embryo itself.

Jellyfish

PHYLUM: Cnidaria CLASS: Scyphozoa ORDER: Semaeostomae

There are about 200 species of true jellyfish. They live in Earth's oceans, from the equator to high latitudes.

Anatomy and taxonomy Scientists categorize all organisms in taxonomic groups based partly on anatomical features. True jellyfish belong to the phylum Cnidaria, or Coelenterata, which also includes hydroids, sea anemones, sea fans, and corals. Additionally, the Cnidaria includes 15 species of sea wasps and box jellyfish. Some hydroids have jellyfish-like stages in their life cycle, but for various reasons are not grouped with the true jellyfish. Comb jellies look superficially like true jellyfish but belong to a completely different group, the phylum Ctenophora.

- **Animals** Jellyfish, like other animals, are multicellular and get the nutrition they need by consuming other organisms. Animals differ from other multicellular life-forms in their ability to move from one place to another (in most cases, using muscles). They generally react rapidly to touch, light, and other stimuli.

- **Ctenophores** Members of the phylum Ctenophora are named for the Greek words for "bearing combs." Ctenophores share many features with members of the phylum Cnidaria. Both are radially symmetrical, though many species are not actually smoothly disk-shaped or cylindrical. They have a two-layered body; they have only one body cavity, the gastrovascular cavity; and their nervous system is simple and not centralized. Unlike cnidarians, however, ctenophores have a continuous gut with a mouth and at least one anus, and they have eight rows of hairlike cilia that run along the length of the body as "combs." These are the structures that give ctenophores their name. The cilia beat in rhythm to propel the animal forward. Ctenophores lack the stinging cells of cnidarians, but have colloblast cells that discharge sticky threads. Most ctenophore species are both male and female at the same time: they are hermaphrodites. About 100 species of ctenophores are now

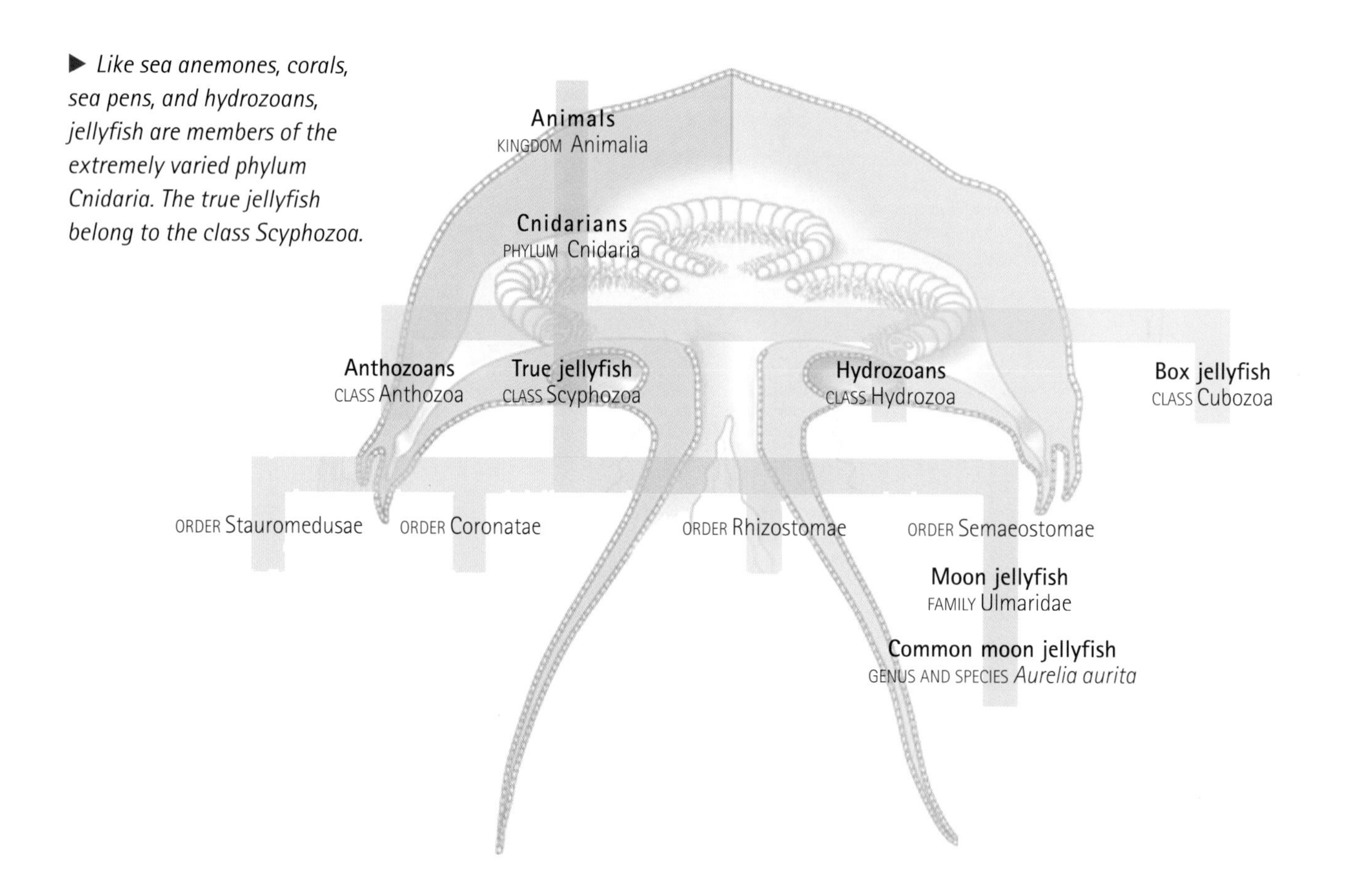

► *Like sea anemones, corals, sea pens, and hydrozoans, jellyfish are members of the extremely varied phylum Cnidaria. The true jellyfish belong to the class Scyphozoa.*

▲ *The lion's mane jellyfish is one of the largest species: its disk-shape bell may be more than 3 feet (1 m) in diameter.*

known. Since these organisms are relatively small and fragile, and some can live at great depths, there are probably many more species as yet undiscovered.

• **Cnidarians** Cnidarians are a very varied group of relatively simple animals. Despite their diversity of form, cnidarians share several major features. They are based on a radially symmetric design, with body parts repeated in a circle around a central axis. They have a simple body design with a body wall consisting of only two layers of cells, between which is sandwiched a gel-like material called mesogloea, the "middle jelly." This arrangement is called diploblastic, meaning "double bud," in reference to the origin of tissues from two cell layers in the embryo. More complex metazoans have three body layers and are described as triploblastic, "triple bud," with tissues originating from three cell layers in the embryo.

Cnidarians have only one major cavity within the body. This is a gastrovascular cavity, or coelenteron, that serves as both a digestive and a transport system. Cnidarians lack the respiratory, circulatory, and excretory organs found in more complex invertebrates. The common feature that gives the Cnidaria their name is the existence of stinging structures (nematocysts) in cells called cnidocytes in their body walls. The word *cnidarian* comes from the Greek for "nettle."

Many cnidarians have two strikingly different body forms in their life cycle. The polyp form is the dominant form in sea anemones, corals, and hydroids. A polyp is essentially a tube with a ring of tentacles surrounding a mouth at one end and a flattened region for attachment at the other end. The other form of a cnidarian's life cycle is called the medusa, after the snake-haired Gorgon of Greek mythology. The medusa is a jellyfish.

• **Hydrozoans** In most of the 3,200 species of hydrozoans, the polyp form is the dominant one, as in the freshwater hydras and the marine genus *Obelia*. The coelenteron is simple and undivided. Gonads develop from the outer cell layer, the epidermis. Hydrozoan medusae have a shelf of tissue called a velum that extends inward from the margin of the bell. A few hydrozoan species, such as the Portuguese man-of-war, are neither polyp nor medusa but make up a floating colony that combines the two forms.

FEATURED SYSTEMS

EXTERNAL ANATOMY The moon jellyfish body is shaped like the bell of an umbrella, with a mouth and oral arms underneath, and tentacles at the rim. *See page 639–640.*

INTERNAL ANATOMY The body wall is essentially two-layered, with a gel-like layer in between. The body wall encloses a single body cavity, the coelenteron or gastrovascular cavity, from which a system of transport canals emerges. The coelenteron and canal system aid the delivery of oxygen and nutrients to body cells, and the removal of waste. *See pages 641–642.*

NERVOUS SYSTEM There is no brain or central nervous system. A double nerve net system carries nerve impulses directly from sensory structures to responding muscle cells. *See pages 643.*

DIGESTIVE AND EXCRETORY SYSTEMS The gastrovascular cavity serves as a gut. Food enters through the mouth and waste material exits through the mouth. There is no anus. There are no structures specialized for waste removal; wastes simply diffuse from body cells into the coelenteron and canal system or directly into the surroundings. *See pages 644–645.*

REPRODUCTIVE SYSTEM A polyp, called a scyphistoma, is the asexual phase of the life cycle, and the medusa is the sexual phase. *See pages 646–647.*

▲ *The bell jelly is a hydrozoan jellyfish. In jellyfish belonging to this class, the base of the bell has a shelf of tissue called a velum.*

• **Anthozoans** The 6,200 or so species of anthozoans include the sea anemones, corals, sea pens, and sea fans. They are all polyps that lack a medusa stage in the life cycle. The polyp's coelenteron is divided by vertical partitions called mesenteries. Some groups have only eight tentacles, but most have six, multiples of six, or simply large numbers of tentacles.

• **Cubozoans** The class Cubozoa consists of the sea wasps and box jellyfish and contains only 15 species. The medusa is the dominant stage in the life cycle. Cubozoan jellyfish have a velum like that of hydrozoans plus a thick mesogloea and gonads that develop from the inner layer, the gastrodermis, as found in scyphozoans. The polyp stage produces a single medusa, which has a boxlike shape and a tentacle at each of the lower corners of the box. The margin may bear up to 24 eyes. Cubozoans are mostly tropical marine species. They are known for their highly venomous sting, which can cause excruciating pain in humans and in some cases death.

• **Scyphozoans** Scyphozoan jellyfish are the "true" jellyfish. Most species have a polyp stage in their life cycle. That stage is usually small and inconspicuous. The polyp stage develops into a form called a scyphistoma that produces many medusae. These bud off the scyphistoma and are the dominant form in the life cycle. Adult medusae can grow very large—more than 6 feet (1.8 m) in diameter with tentacles tens of feet long in some species. The mesogloea makes up the bulk of the jellyfish, and the coelenteron is usually divided by mesenteries into four compartments. The medusae lack a velum. Most jellyfish eat animal plankton (zooplankton), fish, and squid. Some filter the water for microscopic plankton, and a few contain algae that photosynthesize to provide some or all of the animal's nutritional requirements.

• **Semaeostome jellyfish** Semaeostome jellyfish are the most familiar jellyfish of temperate and tropical seas, and include the moon jellies and the lion's mane jellyfish. The mouth typically has four oral arms, and the stomach contains gastric filaments.

• **Rhizostome jellyfish** In rhizostome jellyfish there are eight oral arms, each of which has a secondary mouth. The stomach does not contain gastric filaments as found in semaeostome jellyfish. The "upside-down" jellyfish, *Cassiopea*, is a rhizostome species.

• **Moon jellyfish** The family Ulmaridae, the moon jellyfish, has at least 30 species. The bell of moon jellyfish contains radial canals and a ring canal. In 2003 scientists described a new deepwater species named "big red," *Tiburonia granrojo*, because of its color and size. It grows to more than 3 feet (0.9 m) across. It is so unlike other members of the family—it has four to seven arms and wartlike clusters of stinging cells—that it has been placed in a subfamily of its own.

The common moon jellyfish floats in the surface waters of all oceans between latitudes 70°N and 30°S. It is probably the world's most abundant large jellyfish of the upper ocean. It feeds on small planktonic organisms that it immobilizes using stinging cells.

External anatomy

The jellyfish body, or bell, is shaped like an upside-down saucer or a disk that is indented on the underside. The body parts are arranged in a circle around the central axis, so there is no recognizable front or rear end. The upper side of a jellyfish is the exumbrella; the underside, where the mouth is located, is the subumbrella. Sensory organs are spaced equally around the bell so that the jellyfish can respond to stimuli from any direction. The tentacles can gather food from all directions beneath the bell. Apart from the ring of short tentacles at the edge of the bell, four tentacle-like oral arms extend around the mouth. Like the tentacles, these arms have stinging cells. The oral arms capture food and carry it to the mouth and also transfer food caught by the outer ring of tentacles.

The rim of the bell is separated into eight lobes with the sense organs lying in notches between them. Around the mouth lie four horseshoe-shaped patches that are the gonads, the structures that produce gametes (sex cells). The gonads are red or purple in males and yellow in ripe females. Many of the moon jellyfish's other internal structures are also visible through the semitransparent body wall.

COMPARATIVE ANATOMY

Scyphozoans, hydrozoans, and cubozoans

True jellyfish belong to the class Scyphozoa. They have an umbrella-shaped bell that lacks a velum covering the bell's base. Hydrozoan jellyfish are shaped like an umbrella and have a velum, or they form unusual floating colonies that combine the polyp and medusa forms. Box jellyfish or cubozoans, as their name implies, are box- or cube-shaped. Like hydrozoan jellyfish, they too have a velum. The velum partially closes off the space beneath the bell. This allows the jellyfish to jet-propel itself more rapidly because when the bell contracts, water is forced out of the reduced opening under greater pressure.

▼ Moon jellyfish

The moon jellyfish has a translucent bell through which it is possible to see some of the internal organs. The bell is surrounded with a fringe of stinging tentacles, which stun the microscopic plankton that form this animal's diet.

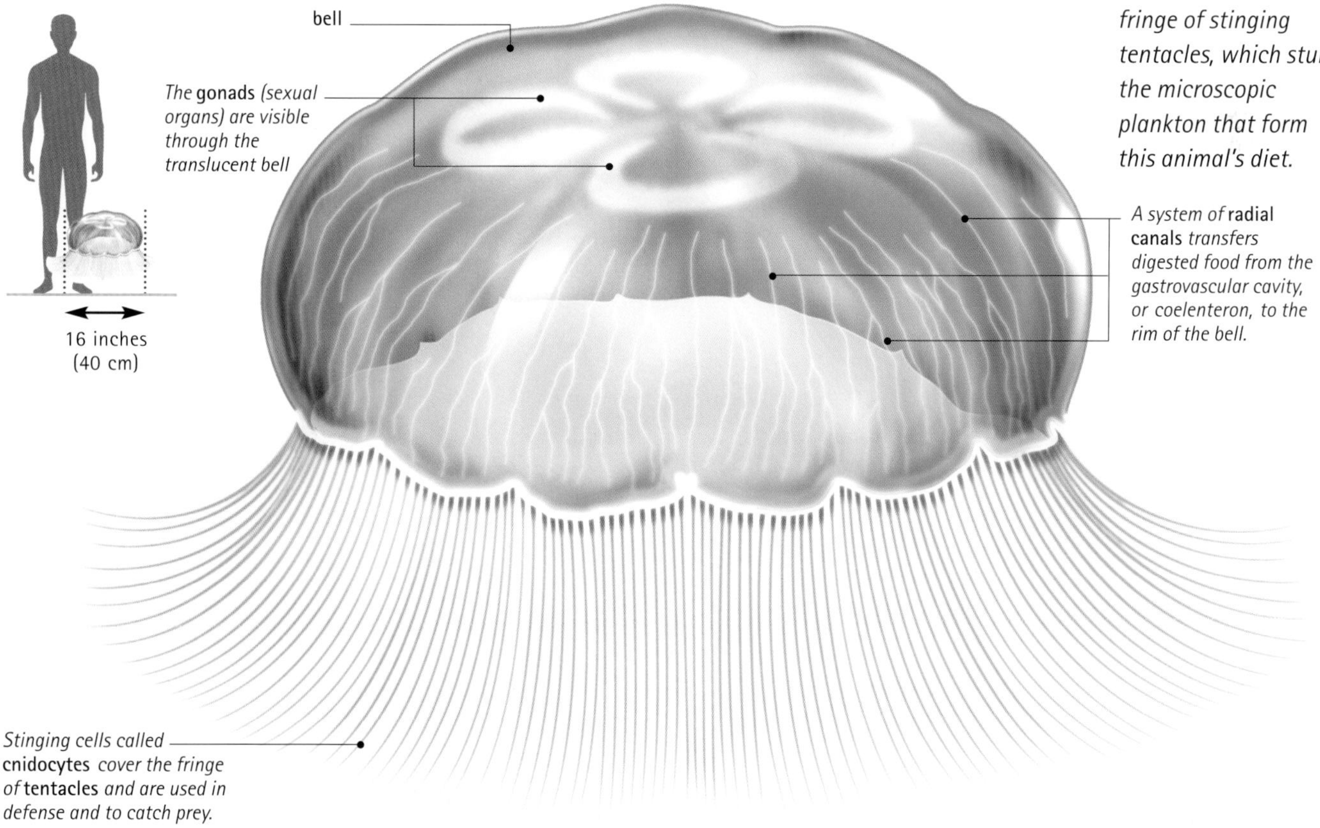

▶ **CNIDOCYTE**
The firing mechanism of the cnidocyte is incredibly rapid. After being triggered, the nematocyst fires within a few thousandths of a second.

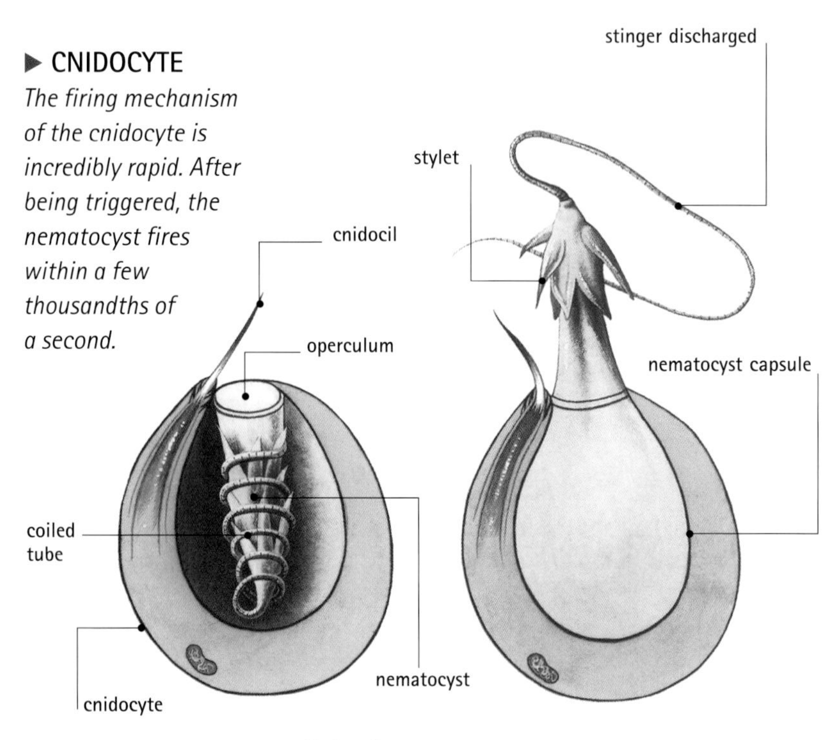

Stinging structures

Like cnidarians in general, the common moon jellyfish uses stinging cells called cnidocytes on its tentacles to capture prey and to defend itself. A cnidocyte starts out in the body wall as a cnidoblast. Inside the cnidoblast, a special internal structure (organelle), called a cnida, is formed. As the cnida matures, the cnidoblast becomes a cnidocyte. Cnidoblasts and cnidocytes are often referred to collectively as nematocysts. Other forms of nematocyst ensnare or entangle prey and attackers, rather than kill them, and can also be used for locomotion and attachment.

The nematocyst is a long, hollow, coiled tube with associated structures such as a trigger, or cnidocil; and a hinged lid, or operculum. The cnidocil is activated by physical contact or by vibrations in the water. Some chemicals—such as those released by injured prey—can also trigger the nematocysts to fire their cnidae. Once triggered, the cnidocil acts like a release catch, causing the operculum to swing open and the coiled tube to discharge. The tube uncoils and turns inside out, injecting venom into a victim's body surface. In some types of nematocysts, the tubule has a hole at the end through which venom is injected. In others, the surface of the tubule is coated with venom, and damage caused to the victim's surface by barbs or similar structures allows venom to enter.

IN FOCUS

Venom

The venoms used by cnidarians such as jellyfish, coral, and sea anemones have to be powerful because these slow-moving animals need to subdue fast-swimming creatures such as fish and shrimplike crustaceans in order to survive. Powerful venom almost instantaneously immobilizes the prey and prevents it from swimming out of reach. The venom of the moon jellyfish is mild relative to that of many jellyfish and poses little or no threat to people. However, the stings of the Portuguese man-of-war (a hydrozoan) and the lion's mane jellyfish (a scyphozoan) are severe enough to cause excruciating pain and some paralysis, which for a swimmer at sea can be life-threatening. By far the biggest threat to humans comes from the tropical box jellyfish of the cnidarian class Cubozoa. Some box jellies are the most venomous creatures in the sea. Brushing against their tentacles triggers nematocysts that inject venom more potent than that of the most venomous snakes. The venom of the box jellyfish *Chironex* contains chemicals that block the action of nerve cells. They can halt the breathing and heartbeat of an adult human within three minutes. Around the world, dozens of people die each year from box jellyfish stings, with many attacks probably going unrecorded.

▶ *The Portuguese-man-of-war is a hydrozoan with powerful stinging cells. This animal's sting can cause temporary paralysis in humans.*

Internal anatomy

The body wall surrounds the coelenteron, or gastrovascular cavity, and has an outer layer of cells, the epidermis; an inner layer, the gastrodermis; and gel-like material in between, the mesogloea.

The epidermis contains sensory cells and stinging cells, the cnidocytes. In addition, gland cells secrete a protective layer of mucus onto the surface of the body, and epitheliomuscular cells can contract to flex the body wall for swimming or to churn the contents of the coelenteron and canal system. Small interstitial cells lying between the other cells are capable of developing into other kinds of cell should any become lost or damaged.

Three main types of cells form most of the gastrodermis. There are mucus gland cells similar to those on the epidermis; enzymatic gland cells that secrete digestive enzymes; and nutritive-muscular cells, which play a dual role. First, they take up digested and partially digested foods; second, they contract to bend the body wall in a manner similar to the epitheliomuscular cells of the epidermis.

Jellyfish and other cnidarians have a simple nervous system that extends as a network of nerve cells at the base of both the epidermis and the gastrodermis. This "double nerve net" is unusual because many of the nerve cells can transmit nerve impulses in both directions. In most animals, nerve cells have a structure and arrangement that permits the transmission of nerve impulses in only one direction. In cnidarians, whenever part of the body wall is stimulated, the electrical signals from that region run out in all directions, often causing that part of the body wall to contract away from or move toward the stimulus.

COMPARE the body wall of a jellyfish with that of a ***SEA ANEMONE*** and an ***EARTHWORM.***

COMPARE the hydrostatic skeleton of a jellyfish with the exoskeleton of an arthropod such as a ***LOBSTER.***

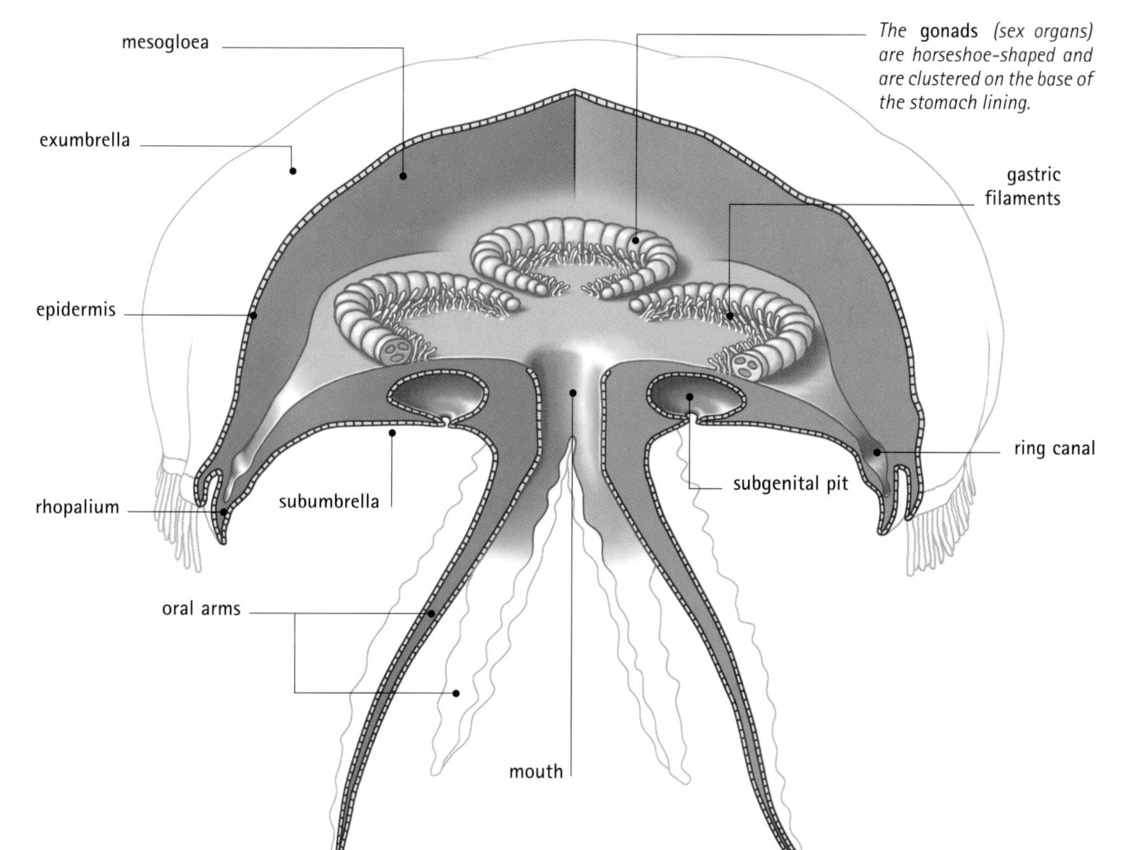

◀ Moon jellyfish

The internal structure of a moon jellyfish is simple in comparison with that of a vertebrate. The yellow color of the gonads indicates that this is a ripe female: that is, she is ready to breed.

▶ **BODY WALL OF A CNIDARIAN**
The body wall of a cnidarian is thin enough to enable respiratory gases to diffuse directly across the cell walls with no need for respiratory organs such as gills. Cells in the body wall each have particular tasks such as defense and the stunning of prey, detecting chemicals in the surrounding water, and producing chemicals called enzymes that speed up chemical reactions.

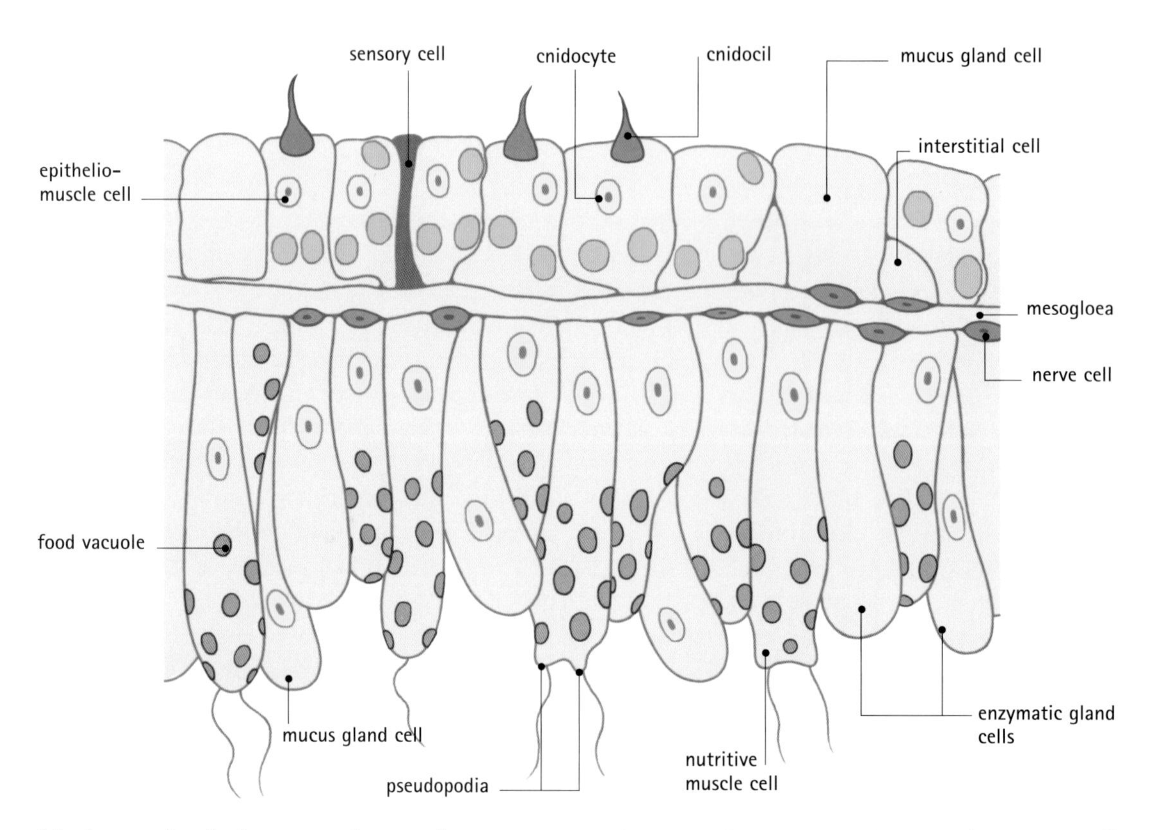

Hydrostatic skeleton and muscles

The jellyfish's seawater-filled coelenteron and gel-like mesogloea form a supportive structure called a hydrostatic skeleton. Both provide some resistance, against which the muscle cells of the bell can contract. The mesogloea also provides support for the body as a whole.

Circular sheets of muscle fibers lie close to the outer edge of the bell and on the underside. They are partly embedded in the mesogloea; and when the fibers contract, they fold the edges of the bell inward and downward in a rhythmic pulse. This forces water out from beneath the bell and pushes the jellyfish upward by weak jet propulsion. When the muscles relax, elastic fibers lying across the mesogloea cause the bell to regain its former shape, ready to contract again.

Most jellyfish alternate swimming upward in the water column with very slowly sinking downward to capture prey that drift or swim into their tentacles. Some jellyfish, including the common moon jelly, are attracted to the surface by the moderate light of a dull day but move away from the surface in bright sunlight or in darkness. By changing direction, and by adjusting the combination of swimming and drifting, moon jellyfish can control their depth in the water. They often gather at specific points in the water column where a small change in depth brings a relatively large change in conditions—for example, where the temperature or salinity changes drastically in a small vertical distance. These zones of rapid change, known as thermoclines (where temperature changes) or haloclines (where salinity changes), sometimes attract larger concentrations of zooplankton, and the moon jellies gather to feed on them.

Combined systems

Jellyfish, like other cnidarians, do not have a separate circulatory system. The coelenteron serves as the circulatory, digestive, and excretory system combined. It circulates partly digested nutrients around the body and takes up waste substances from the surrounding cells. It expels the wastes through the mouth.

Jellyfish, like other cnidarians, lack special organs for gas exchange. The bulk of a jellyfish is largely nonliving material—the gel-like mesogloea. The mesogloea contains very few living cells and, therefore, does not require much oxygen or release much metabolic waste. Overall, any living cells in the jellyfish are found in thin layers or small clusters, so that any one cell is only a short distance from a surface where gases can be exchanged.

Nervous system

Most animals have a centralized nervous system, with collections of nerve cells gathered at the "head" end because this is the part of the body that first samples the environment. Light-detecting structures and chemical sensors are usually concentrated there, and the nervous system that processes sensory data and coordinates responses is concentrated there, too. However, jellyfish, like other cnidarians, do not have a recognizable head or tail end, and, because the jellyfish is arranged on a circular plan, any region of the body could be the first to sample new stimuli. The nervous system is therefore noncentralized, without any structures recognizable as a primitive brain.

Around the edge of the moon jelly's bell lie eight sensory centers called rhopalia (singular, rhopalium). Each contains a pair of chemical-sensitive pits, a balance organ called a statolith, and a light-sensitive structure called an ocellus (plural, ocelli). The pits and ocelli enable the jellyfish to respond to differences in the salinity of seawater and light intensity. The balance organs detect the orientation of the jellyfish's bell relative to the pull of gravity. Taken together, the information from the various sense organs enables the jellyfish to maintain itself in conditions that give it the best chance of survival. The chemosensory pits, ocelli, and statoliths are connected to a complex of nerve cells beneath. If these connections are severely damaged, the jellyfish no longer swims spontaneously. If they are slightly damaged, the jellyfish might swim only slowly.

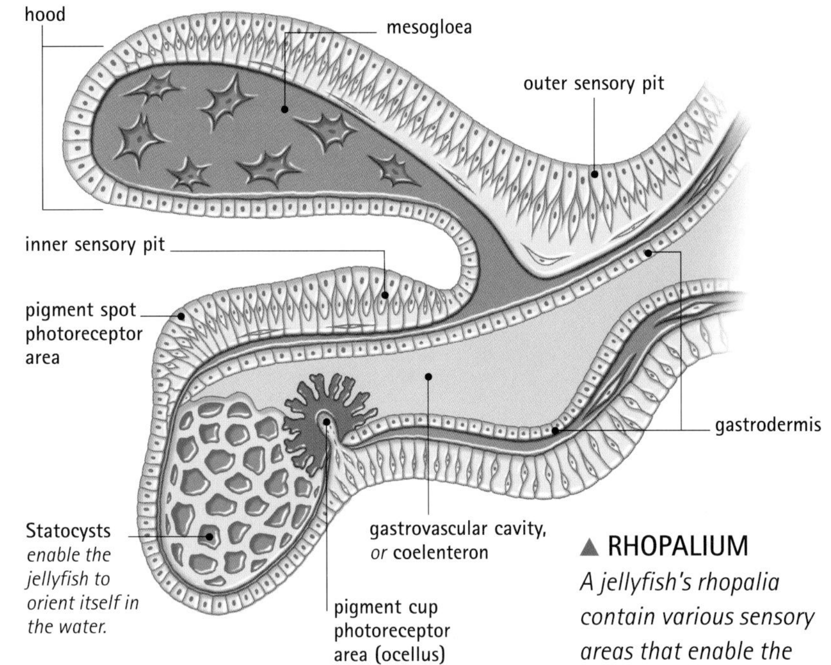

▲ **RHOPALIUM**
A jellyfish's rhopalia contain various sensory areas that enable the animal to detect light and chemicals and to detect the orientation of the bell with respect to gravity.

▼ *The deep-sea dunce-cap jellyfish has four sensory rhopalia, each separated by three tentacles.*

COMPARATIVE ANATOMY

Box jellyfish eyes

Some box jellyfish have up to 24 light-detecting structures or "eyes" that are much more complex than those of other jellyfish. In some species, each eye contains a covering layer or cornea, a round lens, and a light-detecting layer, the retina. These structures are similar to those found in much more complex organisms such as octopuses and fish. Those box jellyfish having complex eyes are usually small, with a squarish body about 1.5 inches (4 cm) long, but they swim fast. Some box jellyfish can travel 200 times their length in one minute. Their highly sensitive eyes enable them to find and intercept the bioluminescent (light-producing) plankton on which they prey.

Digestive and excretory systems

CONNECTIONS

COMPARE the baglike digestive system of a jellyfish with the similar arrangement found in a ***SEA ANEMONE***.

COMPARE phagocytosis in a jellyfish with phagocytosis used by white blood cells in the ***IMMUNE SYSTEM***. In the immune system, the process of phagocytosis is used to destroy invading microbes rather than aid digestion.

Jellyfish, like other cnidarians, do not have a complete gut with a mouth and anus. The cavity in which digestion takes place is more like a bag than a tube. The mouth is both the entrance into which food is taken and the exit through which undigested matter is ejected.

In the moon jellyfish, tentacles at the rim of the bell are armed with stinging cells, or cnidocytes, that capture small zooplankton. Prey items include shrimplike copepods and the larvae of many kinds of invertebrates, including mollusks, crustaceans, and echinoderms, as well as fish eggs and larvae. Oral arms around the mouth scrape the captured animals off the tentacles and carry them to the mouth, where they are swallowed whole. Small projections called gastric filaments project into the coelenteron. The filaments are richly armed with stinging cells. They ensure that any live animals drawn into the coelenteron are quickly killed before they can cause damage.

Digestion happens inside the coelenteron. The lining of the coelenteron, called the gastrodermis, contains cells that secrete digestive enzymes into the cavity. Hairlike structures emerging from the gastrodermis (cilia) help stir the contents of the coelenteron, thus aiding digestion. In the absence of a true circulatory system, the coelenteron is extended into a system of canals. In moon jellyfish, a system of radial canals extends from the gastrovascular cavity to a circular canal that runs close to the rim of the bell. The contents of the system of canals are moved along by the body flexing when the animal swims. This ensures that the coelenteron contents are distributed through the canal system to tissues all around the jellyfish. The coelenteron and canal system ensure that digested food has to travel only a short distance through tissues to reach cells that require nutrients.

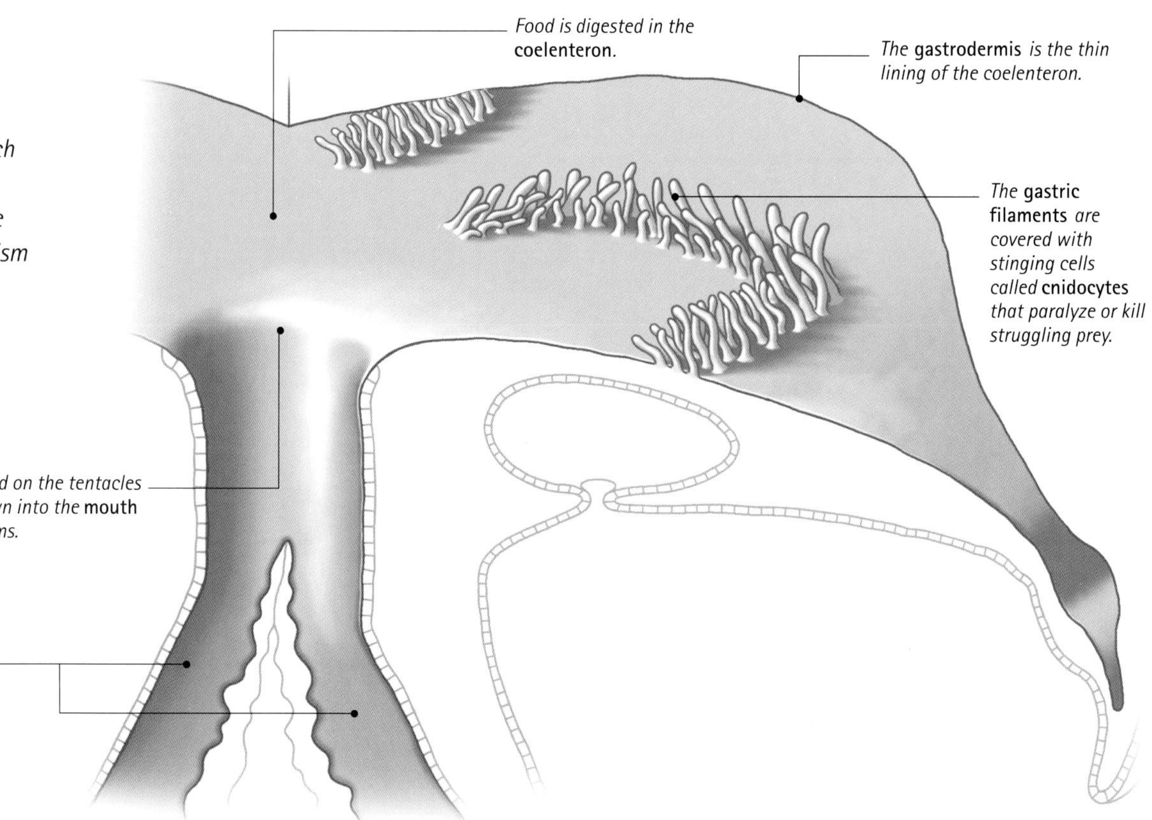

▶ **Moon jellyfish**
A jellyfish's digestive system consists of a single orifice, the mouth, through which food is ingested and waste ejected. Waste products of metabolism are excreted from the surfaces of the jellyfish into the surrounding water.

◀ *The transparent bell of this purple stinger jellyfish reveals its eel prey being digested in the coelenteron.*

After several hours of successful feeding, the gastrovascular cavity contains a soupy broth of broken-down fats, proteins, and carbohydrates from the digested prey. From the coelenteron, the breakdown products are taken up into the nutritive-muscular cells of the gastrodermis. Uptake is done by a combination of two processes. In pinocytosis, or "cell-drinking," the membrane of nutritive-muscular cells folds in on itself and encloses a small droplet of liquid that becomes a vacuole surrounded by a membrane. This increases the surface area across which substances can be absorbed into the cell. In phagocytosis, or "cell-eating," fragments of food are pinched off or engulfed into a vacuole. The cell then secretes digestive enzymes into the vacuoles, where final digestion occurs. Anything that remains undigested in the gastrovascular cavity is simply expelled through the mouth.

Jellyfish, along with other cnidarians, lack specialized excretory organs. Their main nitrogen-containing waste is ammonia, which simply diffuses out of body cells and into the surrounding water. Solid waste is expelled out of the mouth.

IN FOCUS

The upside-down jellyfish

The jellyfish *Cassiopaea*, common in the shallow waters off Florida and in the Caribbean Sea, lies upside down on the seabed. As well as a main mouth, it has many secondary mouths on its oral arms. Small animals captured on the oral arms are wafted by cilia toward one of the many mouths and swallowed. The jellyfish's habit of lying upside down probably relates to its partner algae that live inside the mesogloea. In strong sunlight, the jellyfish can survive solely on the nutrients provided by these photosynthesizing algae.

Reproductive system

CONNECTIONS

COMPARE the life cycle of a jellyfish with that of a *FERN* and *SEAWEED*. All these life-forms have reproductive cycles that involve the alternation of generations, that is, alternating between sexual and asexual phases in the life cycle.

Moon jellies are either male or female. The four horseshoe-shaped gonads—the structures that produce sex cells—lie around the mouth with the opening of the horseshoe directed toward the mouth. In late fall and early winter moon jellies gather in swarms of many thousands. Adult females release eggs and retain them in the gastrovascular cavity close to the mouth. Males release sperm into the seawater, and females use their oral arms to capture the sperm and draw them into the coelenteron where they fertilize the eggs. A fertilized ovum divides many times to form first a tiny hollow ball of cells and later a planula larva consisting of two layers of cells surrounding a fluid-filled center. The female retains her fertilized eggs in her oral arms until they grow into larvae; then she releases them.

The planula larva is about 0.5 inch (12 mm) long and floats in the plankton. Its outer layer of cells has cilia that beat in a synchronized fashion to propel the larva through the water.

GENETICS

Alternation of generations

The life cycle of the moon jellyfish has two alternating phases: the medusa and the polyp, or scyphistoma. Both can reproduce, but only the medusa can produce gametes and reproduce sexually. The polyp is asexual; that is, it produces only clones of itself rather than genetically new offspring.

There are benefits to this complicated arrangement. The larvae that form as a result of sexual reproduction by the medusae are highly mobile—they help the species to disperse over wide areas by drifting perhaps hundreds of miles on ocean currents. Sexual reproduction (mixing of genetic material from two parents) produces genetic variation among the offspring. If environmental conditions change drastically, causing some individuals to die, others are likely to survive, because their particular combination of genes makes them better able to cope with the change. The asexual polyp phase that grows into a scyphistoma and then buds off medusae cannot provide its offspring with this kind of variability, but this does not matter as long as the environment is stable. Circumstances that suit the parent will suit its genetically identical offspring, which can be produced in large numbers to take advantage of the favorable conditions.

The female moon jelly produces **eggs** *in her* **gonads** *and holds them in the coelenteron.* **Sperm** *released into seawater by males lodge in the female's oral arms, which draw the sperm into the gastric cavity, where they fertilize the eggs.*

egg

cilia

The fertilized egg develops into a free-swimming **planula larva.** *It moves through the water by beating its fringe of tiny hairlike* **cilia.**

The planula settles onto a hard surface such as a rock and becomes a **scyphistoma polyp.**

▲ *A tiny ephyra eventually develops into the familiar adult jellyfish form called the medusa.*

After a few days, the larva settles on a rock or shell on the shallow seabed, loses its cilia, and grows into a polyp called a scyphistoma. This looks like a miniature sea anemone.

Over the winter, the scyphistoma grows to about 0.5 inch (12 mm) tall, and in the following spring or summer it undergoes a process called strobilization in which it buds off a series of small jellyfish, called ephyra. Just before they are released, the ephyra are stacked upside down, one on top of another like a pile of saucers. The ephyra has eight distinct arms rather than the continuous bell shape of the adult form. As the ephyra grows, numerous tentacles develop on the rim, and the tissue between the arms grows rapidly so that the ephyra soon assumes a bell shape. Over six to nine months, the ephyra gradually develops the typical structures of the adult. Adult jellyfish usually survive for less than a year. Many die soon after spawning. Others are eaten by turtles, sunfish, and larger jellyfish.

TREVOR DAY

◀ *Several stages are involved in the life cycle of a moon jellyfish. These are the medusa, planula larva, scyphistoma polyp, and ephyra. The egg, planula larva, and ephyra are shown here magnified many times.*

Further Reading and Research

Arai, M. N. 1997. *A Functional Biology of the Scyphozoa.* Chapman and Hall: London.

Brusca, R. C., and G. J. Brusca. 2003. *Invertebrates.* 2nd ed. Sinauer Associates: Sunderland, MA.

Ruppert, E. E., R. S. Fox, and R. B. Barnes. 2004. *Invertebrate Zoology: A Functional Evolutionary Approach.* 7th ed. Brooks Cole Thomson: Belmont, CA.

Kangaroo

ORDER: Diprotodontia SUBORDER: Phalangerida
FAMILY: Macropodidae

Kangaroos are Earth's best-known marsupials and among the most easily recognized of all mammals. They are famous for their spectacular hopping gait and for rearing their young in a special pouch. They have become a symbol of their native country, Australia.

Anatomy and taxonomy

Scientists group all organisms into taxonomic groups based largely on anatomical features. Kangaroos belong to a group of mammals called marsupials, which also includes wombats and possums.

- **Animals** Members of the animal kingdom are multicellular organisms. They are heterotrophic: they obtain energy and nutrition by eating other organisms. Animals are able to move about mostly by using their muscles, and they have a variety of senses through which they are able to respond rapidly to external stimuli.

- **Chordates** Chordates are animals in which the long axis of the body is supported by a stiff rod called the notochord at some stage in the life cycle.

- **Vertebrates** Vertebrates are animals with a bony or cartilagenous backbone consisting of several units called vertebrae. Vertebrate animals have bilateral (mirror) symmetry, a distinct head at the front, and muscles arranged in symmetrical pairs along the length of the body.

- **Mammals** All mammals are warm-blooded; most have fur. In mammals, the lower jaw hinges directly with the skull. Females feed their young on milk secreted by mammary glands. Mammalian red blood cells do not contain nuclei, unlike those of other vertebrate groups.

- **Marsupials** These mammals are an early offshoot within the mammalian family tree. Marsupial females give birth to underdeveloped young, and the offspring complete their development outside the womb, usually in a pouch on the female's abdomen. Marsupials have evolved an enormous diversity of forms, many of which parallel those of the more familiar placental mammals. They include the badgerlike wombat, catlike quolls, squirrel-like possums, and, of course, the kangaroos and their relatives. One of the most diverse orders of marsupials is the Diprotodontia. The name "diprotodont" refers to the characteristic of having only one pair of lower incisors. These animals are also syndactylous—that is, the second and third toes of the hind feet are always fused. Arboreal (tree-climbing) members of the group, such as possums and koalas, usually have an opposable big toe on the hind feet to grasp branches. Ground-dwelling species (kangaroos and wombats) do not have this toe.

Animals
KINGDOM Animalia

Chordates
PHYLUM Chordata

Vertebrates
SUBPHYLUM Vertebrata

Mammals
CLASS Mammalia

Marsupials
SUBCLASS Metatheria

Placental mammals
SUBCLASS Eutheria

Wombats and koalas
SUBORDER Vombatiformes

Kangaroos and possums
SUBORDER Phalangerida

Kangaroos
FAMILY Macropodidae

Possums
FAMILY Phalangeridae

Red kangaroo
GENUS AND SPECIES
Macropus rufus

Eastern gray kangaroo
GENUS AND SPECIES
Macropus giganteus

Wallaroo
GENUS AND SPECIES
Macropus robustus

Red-necked wallaby
GENUS AND SPECIES
Macropus rufogriseus

◀ *This family tree shows the kangaroos' closest relations. The marsupial order Diprotodontia contains 10 families and 131 species. The suborder Phalangerida is made up of several families, including the Macropodidae—the kangaroos and wallabies.*

• **Phalangerida** This large group contains all the long-tailed members of the diprotodonts. It includes four groups of possums and the kangaroos.

• **Macropodoidea** The Macropodoidea superfamily unites the kangaroos and wallabies (Macropodidae) and their close cousins, the Potoroidae. These latter include the primitive rat-kangaroos, potoroos, and bettongs. The Potoroidae are all small, brown, jumping marsupials with a thin, ratlike tail, rather like a small rodent. Potoroids feed on fruits, nuts, insects, and other invertebrates.

• **Kangaroos and wallabies** The Macropodidae family is made up of 12 genera and 61 species of kangaroos and wallabies. They have long hind feet, in which the fourth and fifth toes bear the animals' weight. The fused third and second toes are smaller, and the first toe is absent. The hind legs are large, and the forelegs are small and armlike, each with five digits. Macropods live in both arid and temperate grasslands, and in rocky outcrops and tropical forest.

Among the Macropodidae, the rock wallabies (in the genus *Petrogale*) are the most agile of all the kangaroos, and they can travel fast over precarious boulder slopes and rocky outcrops. Several are very colorful, such as the yellow-footed rock wallaby, which has a banded tail and facial markings. Tree kangaroos (in the genus *Dendrolagus*) have returned to the arboreal lifestyle of their ancestors. Tree kangaroos' hind feet are shorter than those of ground-dwelling kangaroos and have soft, flexible pads that help them grip branches. The tail is long and furry, and acts as a counterbalance when the animal is climbing. The quokka is a secretive species of wallaby, little bigger than a hare. The quokka lives only in the extreme southwestern region of Western Australia.

▲ *The red kangaroo is the largest of the kangaroo species. It has the hopping gait and marsupial pouch typical of all kangaroos.*

• ***Macropus* kangaroos** There are four large and widespread species in the genus *Macropus*: red, eastern and western gray, and the wallaroo. The western gray kangaroo is generally smaller and browner than its eastern cousin. The genus also includes several smaller species, called wallabies, two of which weigh only 7 pounds (3.5 kg) or less.

FEATURED SYSTEMS

EXTERNAL ANATOMY Kangaroos are medium to large mammals with massively developed hind legs and a long, muscular, tapering tail. *See pages 650–653.*

SKELETAL SYSTEM The skeleton has large bones in the hind feet and legs. The bones of the second and third hind toes are fused by skin. *See pages 654–655.*

MUSCULAR SYSTEM The muscular system of kangaroos is dominated by powerful hind leg muscles and long tendons. These allow kangaroos to hop at high speeds for long periods using very little energy. *See page 656.*

NERVOUS SYSTEM Kangaroos have acute senses of sight, smell, and hearing. *See page 657.*

CIRCULATORY AND RESPIRATORY SYSTEMS Although a kangaroo's heart is small, it is efficient. The peripheral circulation plays an important role in keeping the animal cool. *See page 658.*

DIGESTIVE AND EXCRETORY SYSTEMS The digestive system is highly efficient, with a large, chambered stomach and long intestines. *See pages 659–660.*

REPRODUCTIVE SYSTEM Female kangaroos have two wombs and two vaginas for mating. Young are born as embryos through a third, central vagina. Development is completed in a pouch, where the embryo attaches to a teat. *See pages 661–663.*

External anatomy

CONNECTIONS

COMPARE a kangaroo's hare-lip with that of a ***HARE***.

COMPARE the kangaroo's long neck and ears and sideways-facing eyes with those of a ***RED DEER***. These animals have evolved similar forms for avoiding predation.

The kangaroos and large wallabies are unmistakable. Even someone who has never before seen one of these extraordinary Australian mammals would have no difficulty identifying the fleet, bounding form.

Australia has no native hoofed mammals, and kangaroos have evolved to fit similar types of environment as some nonmarsupial herbivores (plant-eating animals), such as cattle and deer, in other dry parts of the world. All these animals have a similar diet and digestive physiology. The evolution of both groups has also led to similar adaptations for avoiding predation. Both kangaroos and antelopes are tall and alert, with a long neck, large swiveling ears, and eyes located on the sides of the head that offer all-around vision. Kangaroos and antelopes both have long legs, and when alarmed take flight in leaps and bounds. Kangaroos that inhabit open grassland (such as

The **ears** *are large and flexible. They are able to detect distant sounds—for example, the footsteps of an approaching predator.*

In males the **fur** *is russet to brick red on the back, and paler on the throat, belly, and limbs.*

Large **eyes** *provide good night vision.*

The **nose** *is very sensitive to odors. This sense is important for kangaroos, which are most active at night.*

The **forelimbs** *have five digits and are used to manipulate food and in slow locomotion.*

The **hind limbs** *are much larger than the forearms.*

foot

▶ **Red kangaroo**
The body shape of the kangaroo makes it one of the most recognizable of all mammals, with its upright (or hopping) stance, huge hind legs and tail, very short forelimbs, and large, pointed ears.

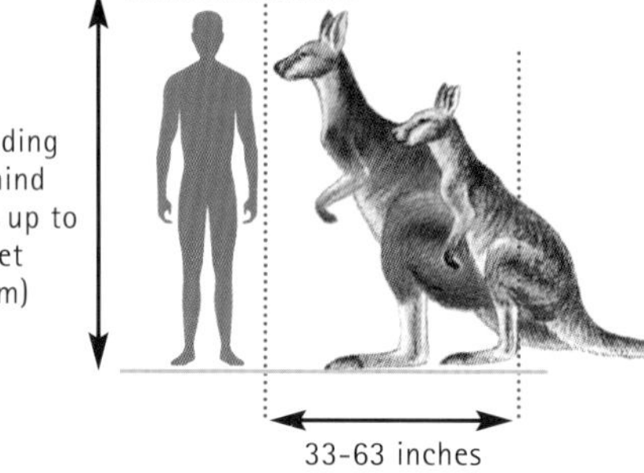

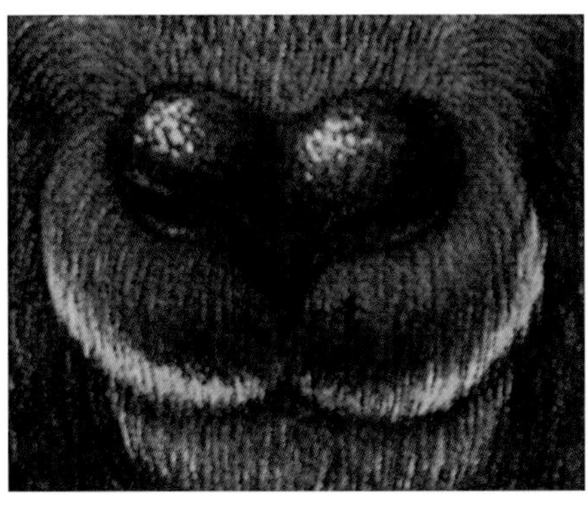

Wallaroo

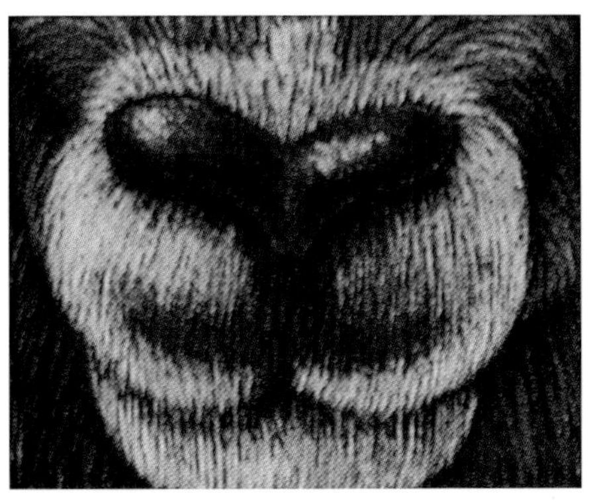

Red kangaroo

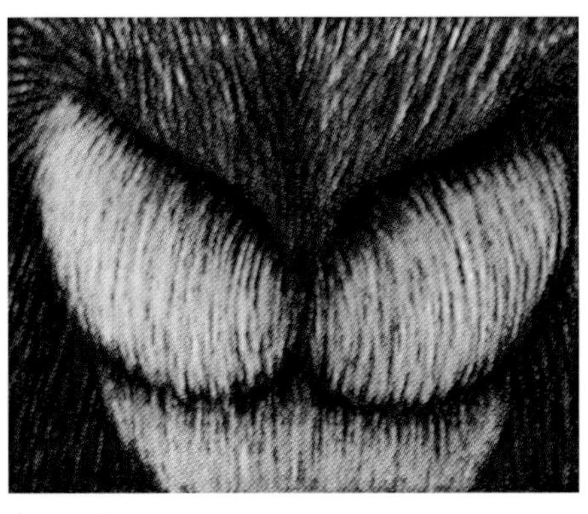

Gray kangaroo

▲ NOSES

The nose area of these three species is very distinctive. The bare area of a wallaroo's snout is black and hairless. The gray kangaroo has the hairiest snout; and the red kangaroo's nose is intermediate.

red kangaroos) live in groups that enhance their awareness of predators. Those species that live in forests, such as the forest wallabies, tend to be solitary.

In both wallabies and kangaroos, the neck is long and slender. The head is small, with large, erect ears and bulging eyes located on the side of the head. The muzzle is blunt and soft, with large nostrils and a harelip (divided upper lip). A small area at the tip of the nose of the red kangaroo is hairless, whereas that of the gray kangaroo is furry. In the wallaroo—a species superficially similar to the gray kangaroos—the snout is naked over a much larger area.

The thick, muscular **tail** *acts as a counterbalance when the kangaroo is bounding and as a stabilizer when the animal is feeding.*

The **hindquarters** *are large and strong.*

▼ *Male red kangaroos can be substantially larger than females. Their fur color is also very different: the male's fur is reddish brown and the female's is bluish gray.*

▼ FOOT, FROM BELOW

Kangaroo

The long, narrow shape of a kangaroo's foot helps give the animal stability when it stands upright and acts like a spring when it lands between bounds. The first digit is absent in the foot, and the small second and third digits are held together by skin. The fourth and fifth digits are strong.

fourth digit

fifth digit

There is a **double claw** *at the end of the second and third digits.*

◀ *The kangaroo hop is a very efficient way of moving around. Red kangaroos can make bounds up to 30 feet (9 m) long or more and reach speeds of up to 30 miles per hour (50 km/h). The smaller gray kangaroo has been recorded jumping 44 feet (13.5 m).*

IN FOCUS

Time dwarfs

Studies of the remains of long-dead marsupials in Australia have revealed that the average size of bones and teeth in most large species has decreased quite dramatically over the last 40,000 years. This strange phenomenon is especially apparent in the large kangaroos, which now appear to be about 30 percent smaller than their ancestors. Why have they shrunk? Scientists believe the reason for this has a lot to do with humans, since the shrinkage began at around the time the first people arrived in Australia. Human hunters tend to target large individuals, so hunting may have exterminated the larger species but left the smaller kangaroos.

Body and tail

A kangaroo's body is large, with a deep but narrow chest and pronounced collarbones. The rump tapers into an enormous tail, which is at least as long as the body. The body itself is covered in soft, slightly woolly fur, the color of which varies considerably within and between species. In the red kangaroo, most males (boomers) are reddish brown with a pale color on their underside, while females (blue fliers) are a shade of bluish gray. Both the male and the female colors blend well with the arid scrub habitat, providing camouflage. The reddish fur of the males matches the color of the poor desert soils, and females can be very difficult to see among the blue-green foliage of shrubs such as saltbush.

EVOLUTION

Kangaroo ancestors

The ancestors of all kangaroos and wallabies were small, forest-dwelling marsupials that fed mostly on fruits, nuts, insects, and leaves, much as the smaller forest wallabies do today. As the grassland habitat opened up, grazing and browsing animals evolved, and some of them grew very large. An extinct species of the Pleistocene epoch (1.6 to 0.01 million years ago), *Procoptodon goliah,* was the largest species of kangaroo ever to have lived. It weighed up to 400 pounds (200 kg), had one well-developed toe on each hind foot, and had unusual front paws with two long and three short digits on each. The forearms were long, and the face was very short, with a deep, powerful-looking jaw.

▶ *This black-footed rock wallaby is a close cousin of the kangaroos. Although considerably smaller, it shares the same distinctive body shape. Its home is among rocky hills and cliffs, and it feeds mostly on grass, although sometimes on bark and roots.*

Male kangaroos are often considerably larger than females; a male red kangaroo can be up to four times larger than a female. Males continue to grow and change body shape throughout their life; old males may not get any taller, but the chest continues to broaden and the forearms become increasingly muscular. For a female, these features are probably indicators that a male has had a long life and would probably be a good choice of mate.

Female kangaroos have a large pouch on the lower part of their abdomen in which they carry their young. All pouched animals, or marsupials, are named for the scientific term for this pouch, the marsupium. The pouch of kangaroos opens toward the front, so there is little danger that the passenger will fall out as its mother stands upright or moves around. The opening of the pouch is very elastic. It closes to a narrow slit but stretches wide to allow older, larger joeys—or kangaroo infants—to clamber in and out.

Kangaroo movement

Like humans, kangaroos are bipedal; they move on two legs. However, kangaroos do not walk (transfer their body weight from foot to foot) as we do, and they cannot move backward. Instead, they bounce on greatly enlarged hind legs. The genus name *Macropus* means "great-foot." The hind feet of the large kangaroo species are narrow, but up to 14 inches (36 cm) in length, with hairless soles covered in hard, calloused skin. The forelegs are small in comparison, especially those of females. The forepaws have five separate digits ("fingers"), each with a long, blunt claw.

Kangaroos are famous for bounding, but they also have another means of getting around. When moving slowly—for example, while grazing—they use a unique "five-legged" technique. In this method of movement, the tail acts as a fifth limb, supporting the back end of the animal while it leans on its forelegs. The hind legs are then swung forward together.

CLOSE-UP

Quokka climbers

Kangaroos and wallabies use their tail as a fifth leg when walking slowly. Their relative the quokka, however, uses only its legs. When moving quickly, the quokka hops in the same way as other kangaroos, but unlike most other macropodids it is also capable of climbing to 5 feet (1.5m) above ground to reach twigs.

Skeletal system

CONNECTIONS

COMPARE the long hind leg bones of a kangaroo with those of other hopping animals such as the ***HARE***.

COMPARE the epipubic bones of a kangaroo with those of a monotreme such as a ***PLATYPUS***.

Kangaroos have large, heavy bones in the hind legs and pelvis, and smaller, more delicate ones toward the front of the body.

The backbone comprises 49 to 53 vertebrae (depending on the individual), of which 7 are in the neck and 21 to 25 make up the tail. The 13 dorsal (back) vertebrae have long processes (projections) pointing up from the spine. These serve as attachment points for the muscles of the front quarters. They are often larger in males, which have bigger shoulder and forearm muscles. The clavicle (collarbone) is well developed in both sexes. The humerus (upper arm) is short, about half the length of the lower arm bones (radius and ulna).

The vertebrae of the lower back (lumbar vertebrae) are large. Bipedal hopping puts large strains on the lower back and the pelvic girdle.

The pelvis has two bony processes called epipubic bones. These were once called the marsupial bones because it was believed that they had evolved to support the weight of the young in the pouch. In fact, they are vestiges (remains) of bones more commonly seen in reptiles, from which the first mammals evolved millions of years ago. Epipubic bones are seen in most marsupials and in monotremes (egg-laying mammals) such as the platypus. Placental mammals probably lost these bones through evolution after they split from the marsupials.

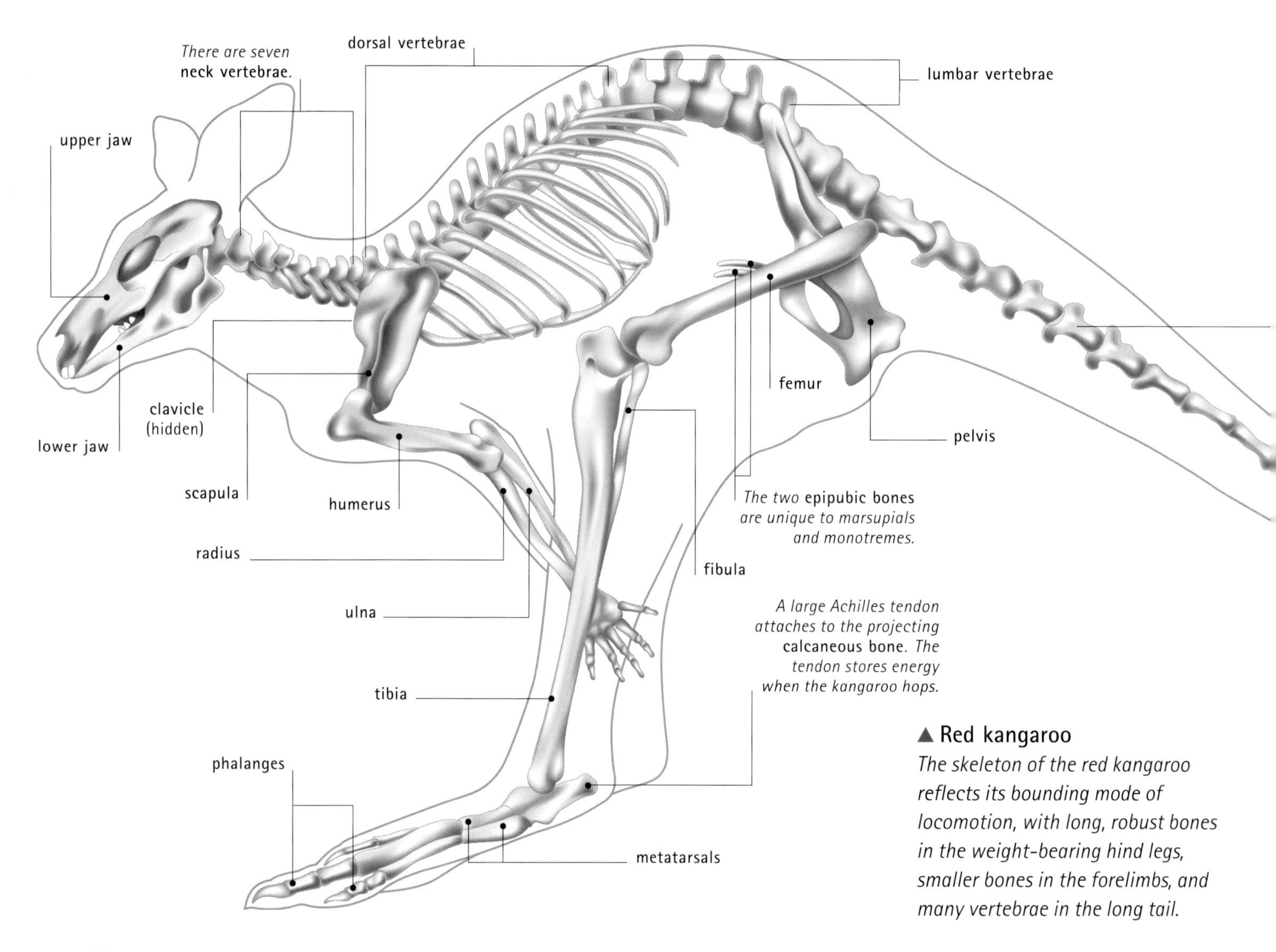

▲ Red kangaroo

The skeleton of the red kangaroo reflects its bounding mode of locomotion, with long, robust bones in the weight-bearing hind legs, smaller bones in the forelimbs, and many vertebrae in the long tail.

▶ PELVIC BONE
Red kangaroo
This bone joins the bones of the hind legs to those of the spine. The two epipubic bones, which survive from the mammals' reptilian past, are not present in placental mammals.

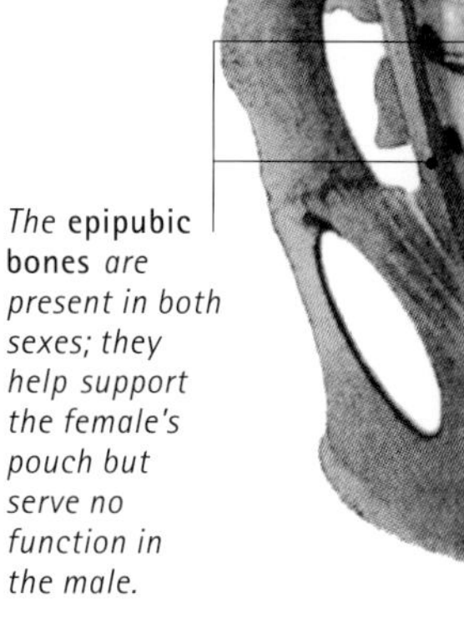

In kangaroos, the thighbone (femur) is short, but all the lower leg and foot bones are very long. The tibia, fibula, and metatarsals in the hind legs are elongated, contributing greatly to the length of the lower leg and the amount of ground that can be covered in a single bound. Kangaroos have no kneecap (patella). The structure of the ankle prevents the foot rotating sideways. Kangaroos, along with possums, wombats, and koalas, are syndactyl animals. The term "syndactyl" means that they have fused, or joined, toes. The second and third toes on the hind feet are bound together by skin to form a single digit with a double claw at the tip. The closely spaced claws in the fused toes make it an ideal tool for grooming fur.

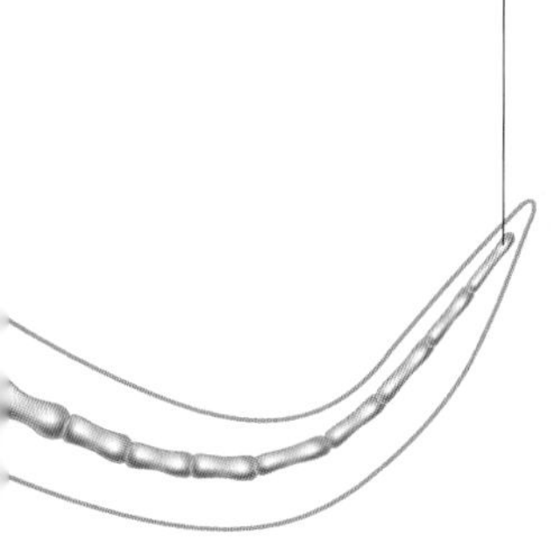

A small skull

The skull of a kangaroo is small and delicate, and the bones covering the braincase are very thin; this explains why kangaroos are killed easily by a blow to the head. The braincase is small, and the palate (roof of the mouth) is incomplete, with spaces in the bone.

The articulation (joints) of the kangaroo jaw allows side-to-side movements for grinding, as well as opening and shutting movements for biting and munching. This arrangement enables plant material to be very thoroughly chewed and ground up before it is swallowed.

CLOSE-UP

Jaws and teeth

Early studies of kangaroo jaws showed that the two sides of the lower jaw are not fused (joined) together. Therefore, they can be pulled a little way apart, separating the two front incisor teeth. Biologists once believed that this separation allowed the lower incisors to work against each other like scissor blades, snipping leaves of grass between their inner edges. However, in the late 1950s, the teeth were examined more closely and it was discovered that they work very much like those of other grazing mammals.

Kangaroos have three pairs of incisor teeth in the upper jaw and just one pair of incisors in the lower jaw. This arrangement is characteristic of all members of the order Diprotodontia (animals with two first teeth), to which kangaroos belong.

In the lower jaw, the front teeth do not line up exactly with those in the upper jaw. This enables the front teeth to escape wear and tear when the animal uses sideways chewing movements to grind tough plant material between its millstonelike molar teeth farther back in the cheek. The front teeth are separated from the cheek teeth by a large gap called the diastema. There are no canine teeth in the lower jaw, but the upper jaw has a pair of canines, which are sometimes absent, as well as its three pairs of incisors.

▼ Skull and jaw
Like all members of the order Diprotodontia, kangaroos have two incisors or "first teeth" on their lower jaw.

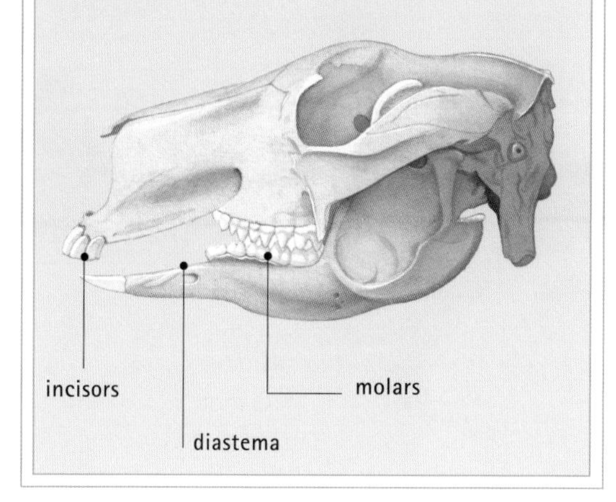

Muscular system

Not surprisingly, the most muscular parts of a kangaroo are its hind legs. The muscles of the hind legs are enormous and are attached to the bone with very strong elastic tendons. The muscles greatly enhance the power of the kangaroo's hopping gait. A kangaroo's tail is also highly muscular and strong enough to support the animal's whole weight. A number of large tendons connect the muscles of the tail to the hip bones.

Many other animals adopt a bounding gait when moving fast, but none manage to use it as efficiently as a kangaroo. One reason for this is that a bounding kangaroo does not use its front legs for support. The hind legs of rabbits, for example, are strong and wonderfully springy, but the momentum they generate is diminished every time the front feet come down; in effect, the front feet act as brakes. Hopping on the back legs alone, a large red kangaroo can travel at up to 30 miles per hour (about 50 km/h) and cover 30 feet (9 m) with each bound. The forelegs do not touch the ground, and the tail is held out behind the animal, acting as a counterweight to prevent it from pitching forward onto its face.

The structure of the hind legs makes the leaps appear almost effortless. Hopping on two legs is a very efficient means of getting around—after the first few hops. These hops use a lot of energy, but once the movement gets going it becomes almost self-sustaining, because the tendons in the animal's legs and tail act like the springs in a pogo stick.

CLOSE-UP

Stretchy tendons

Tendons are long, strong, elastic tissues that connect muscles to bones. When a muscle contracts, the tendon is stretched like a rubber band. The energy stored in a stretched tendon helps the muscles, and the joints they control, to spring back into their original positions.

The tendons in a kangaroo's hind legs are huge. There are also tendons connecting the muscles of the tail to the hip bone. When the animal lands, the legs and tail bend, and energy is absorbed as the tendons stretch. As soon as the animal begins to push off from the ground again, the tendons spring back to their original length, catapulting the whole of the animal's body back into the air.

Since large kangaroos can outrun most potential dangers, they are able to live relatively safely in open environments. It is no accident that quokkas and bettongs, which have much smaller hind legs and hop less often, live in more enclosed habitats of forest or scrub.

The massive muscles in the thighs and tail are not used only for moving around. Mature male kangaroos use them for fighting each other, and females use them in self-defense or to protect their young. Red kangaroos are normally placid but sometimes kill wild dogs called dingoes with a single vicious kick, and they may also sometimes attack humans. When fighting one another, male kangaroos lock their forelegs and wrestle. Each tries to unbalance his opponent, then leans back onto his tail so that the hind legs can swing forward to land a punishing double blow.

▲ **Hop to it!**
As they bound along on their hind legs, kangaroos use their tail as a counterweight in their high-speed forward motion.

Nervous system

Kangaroos have a small brain relative to their body size. On average, the brain is only 2 or 3 inches (5 to 7 cm) long and weighs about 2 ounces (60 g), thus representing only around 0.1 percent of the animal's body weight. The two cerebral hemispheres are considerably smaller than those in a placental mammal of similar size. The cerebral hemispheres contain the cerebral cortex. This part of the brain is characteristically large in most mammals and is concerned with higher functions such as thinking and memory, as well as with vision and hearing.

Like the brain of monotremes (egg-laying mammals), that of marsupials, including kangaroos, lacks a corpus callosum. This is the brain tissue which, in placental mammals, links the two hemispheres of the brain. The corpus callosum is important in allowing information to cross from one side of the brain to the other, uniting the two hemispheres as a single brain. The cerebral hemispheres of diprodonts are connected by the fasciculus aberrans.

The marsupial means of reproduction places unusual requirements on the developing nervous system. Kangaroo joeys leave the womb at a very early stage, compared with placental mammals, and then continue to develop in the mother's pouch. At a correspondingly early stage, the kangaroo embryo develops a rudimentary sense of smell and gravity, so that once it is born it can find its way to the pouch, which lies above the birth canal.

Adult kangaroos have good eyesight and good night vision. Scent is important in social behavior, and males often have a reddish stain on the chest where secretions from a scent-producing gland leak onto the fur.

IN FOCUS

Kangaroo communication

Kangaroos are generally silent, but they will give short, harsh barks when angry or alarmed. Female kangaroos make clucking sounds to call their joeys to their side, and most species of kangaroo use an urgent drumming or thumping of the hind feet on the ground to signal danger.

CONNECTIONS

COMPARE the brain of a kangaroo with that of a placental mammal, such as a ***HUMAN*** or ***DOLPHIN***. A kangaroo's brain lacks the corpus callosum, which connects the hemispheres in the brain of a placental mammal.

▼ BRAIN VIEWED FROM UNDERSIDE
Gray kangaroo

▼ BRAIN VIEWED FROM SIDE
Gray kangaroo

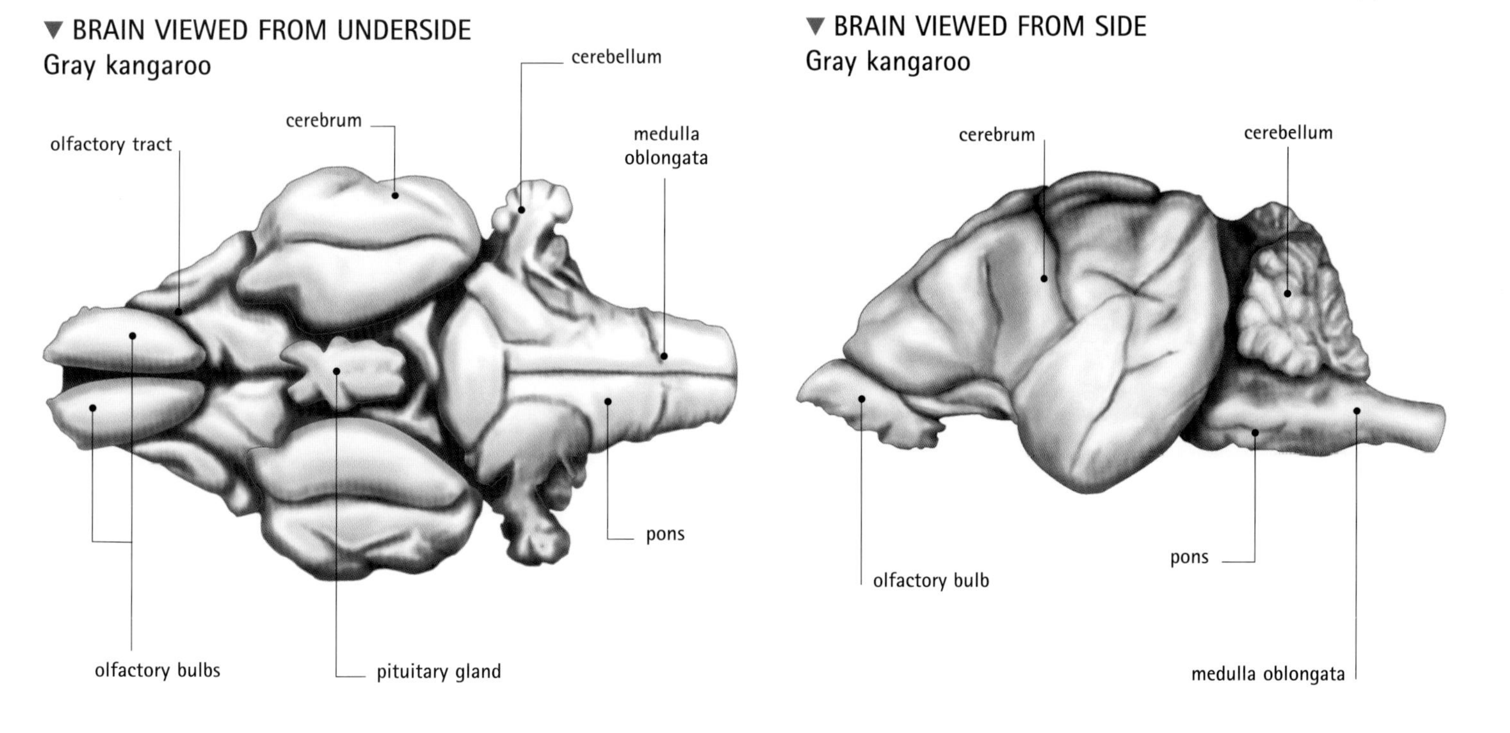

Circulatory and respiratory systems

Despite their many differences from placental mammals, kangaroos and other marsupials have a similar circulatory system. A red kangaroo's heart is about the same size as that of a human. Kangaroos have a closed circulatory system with blood flowing away from the heart in arteries and toward the heart in veins. The blood receives oxygen as it passes through the lungs, and the oxygen is then transported in the blood by red blood cells.

Sweating

Sweating freely as a means of losing heat involves considerable water loss and the risk of dehydration. For kangaroos living in the arid Australian bush, sweating could create more problems than it solves, so kangaroos avoid it when possible. Interestingly, they sweat only while they are hopping. On the move, air passing over the body is an effective means of dissipating the heat generated by exercise, but once the animal comes to a stop, sweating is wasteful. Instead, kangaroos find other means of cooling off. Some of these are behavioral, for example, wallowing in pools, lying the shade, and coming out to feed after dark. In addition the peripheral circulation plays an important part in dissipating heat. The skin of the forearms is very well supplied with blood vessels. By licking its forearms and covering them in a film of moist saliva, the kangaroo speeds up the rate of heat loss. This behavior still involves some loss of water from the body, but at a more controllable rate than would occur with involuntary sweating.

IN FOCUS

Panting

Kangaroos use panting as a way of keeping cool. By breathing fast, they keep a flow of air moving over the moist lining of the windpipe and lungs. Air that has become saturated with water in the lungs and throat is continually replaced with dry air, so that the kangaroo maximizes the rate of evaporation and thus the rate of cooling.

As is the case with other marsupials, the average body temperature of kangaroos is slightly lower than that of placental mammals. No one is really sure why this should be, but it may be that, by running at a lower temperature, the kangaroo uses less energy and is better able to cope with food shortages and other hardships.

▼ *Kangaroos live in dry conditions and need to minimize the need for sweating, so they seek shady places to rest.*

Digestive and excretory systems

The large kangaroos eat mainly grass. In contrast, their ancestors were forest-dwelling animals that fed on fruits, tubers, and succulent leaves. As Australia's climate became hotter and drier between 10 million and 2 million years ago, the forest began to be replaced by grassland and scrub. After a time, certain species of kangaroo ancestors began to specialize in converting this low-grade but abundant vegetable matter into useful sugars and proteins. They were the first and only large, native animals in Australia that were grazers, and thus they flourished.

Because grass is low in energy and nutrients, kangaroos have to eat plenty of it to sustain themselves. The kangaroo stomach is very large. When full, it can account for more than one-seventh of the animal's body weight. However, the size of the stomach is misleading. It allows kangaroos to eat large quantities when they get the opportunity to do so, but on average a kangaroo eats less than similar-size herbivorous placental mammals, such as cattle. A kangaroo's digestion is far more efficient, and it will extract every last ounce of nutrition from its food, while a sheep or cow relies on food's being continuously available, processing large quantities rapidly and less thoroughly than the kangaroo.

In addition to its large size, the kangaroo's stomach has a structure that well suits the animal's diet. Far from being a simple bag, as in

CONNECTIONS

COMPARE the digestive system of a grass-eating kangaroo with that of another grassland herbivore such as a ***RED DEER*** or a ***WILDEBEEST.***

COMPARATIVE ANATOMY

Divided stomachs

Unlike humans, cows and kangaroos both have divided stomachs, which they use to break down the large amount of cellulose in their grass-based diet. In kangaroos, grass is first thoroughly chewed and then stored in the mid stomach, where bacteria break down the cellulose. In contrast, cows are ruminant herbivores, so after food has been partially broken down in the rumen area of the stomach, it is returned to the mouth for further chewing.

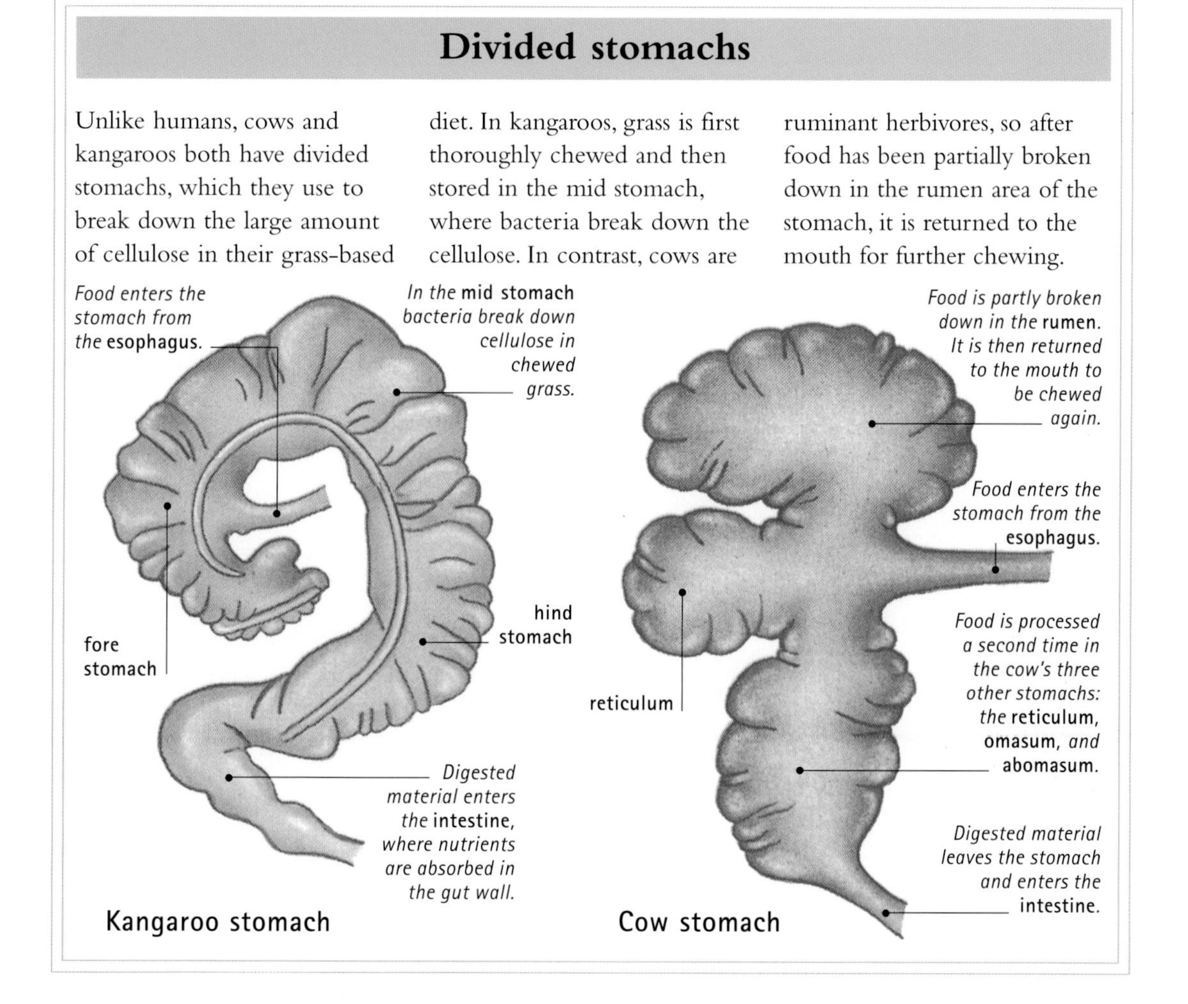

IN FOCUS

Careful chewing

Kangaroos spend a very long time chewing their food and turning it into a fine pulp before swallowing. Unlike cows, they do not regurgitate food to chew cud, so they get only one opportunity to reduce the plant fibers to an easily digested mush.

humans, it is multichambered or "sacculated." Food passes slowly though the different stomach compartments, which are separated by bands of muscles that pucker the lining at intervals. The elongated middle section of the stomach acts as a fermentation chamber, where symbiotic cellulose-digesting bacteria get to work breaking down the tough plant material and converting it into sugars and other easily absorbed compounds. Like most herbivores, kangaroos have a very long intestine to allow plenty of time for full absorption of the products of digestion to take place.

Red kangaroos that have the chance to feed on lush, green grass do less well than those that eat dry, shriveled grass. Because of the large amount of water in fresh grass, weight for weight it contains less energy than dry grass and takes up more stomach space. A kangaroo that eats its fill of dry grass will benefit more than one that fills up on the fresh version.

Kangaroos living in arid habitats can go for weeks without drinking. Instead, they lick dew that forms on leaves or rocks and use all the available moisture in their food. They conserve water by producing concentrated urine, and they avoid the need to sweat by feeding at dawn and dusk and resting in the shade during the heat of the day.

▶ *Unlike ground-dwelling kangaroos, which live on grass, tree-dwelling kangaroos include plenty of leaves and fruit in their diet.*

Reproductive system

The reproductive system of kangaroos and other marsupials is substantially different from that of placental mammals. Female kangaroos have not only two ovaries (as do placental mammals), but also two wombs—whereas in a placental mammal there is just one womb (uterus). The kangaroo also has two long, curved vaginas through which the male's sperm passes on its way to fertilize the eggs. As in all male marsupials, the male kangaroo's scrotum is positioned in front of the penis. When the female is ready to give birth, a third opening develops between the two side vaginas. This birth canal is similar to the single vagina of placental mammals and opens adjacent to the digestive tract in the cloaca. In most marsupials it seals over again after each litter is born, but in kangaroos it becomes a permanent structure after the first birth.

As with all marsupials, young kangaroos are born in an embryonic state. While in the womb, the embryo is surrounded by a thin membrane secreted by the uterus lining, in a manner similar to the eggshell in a reptile. Inside the membrane, the embryo is nourished by its own yolk. The beginnings of a placenta start to form as the yolk runs out, but this never develops fully, because the baby is born soon after, still at the embryo stage. The membrane and fluids in which the embryo develops are born with it and are usually eaten by the mother.

Unlike placental mammals, baby kangaroos complete most of their development outside the womb and do not benefit from a placental link with the mother. Instead, they attach to a teat within an hour or so of being born. The teat provides them with nourishing milk from a mammary gland. Once the baby has latched on, the teat swells inside its mouth so that it does not have to exert any energy to hold on. It will remain attached to the teat for weeks.

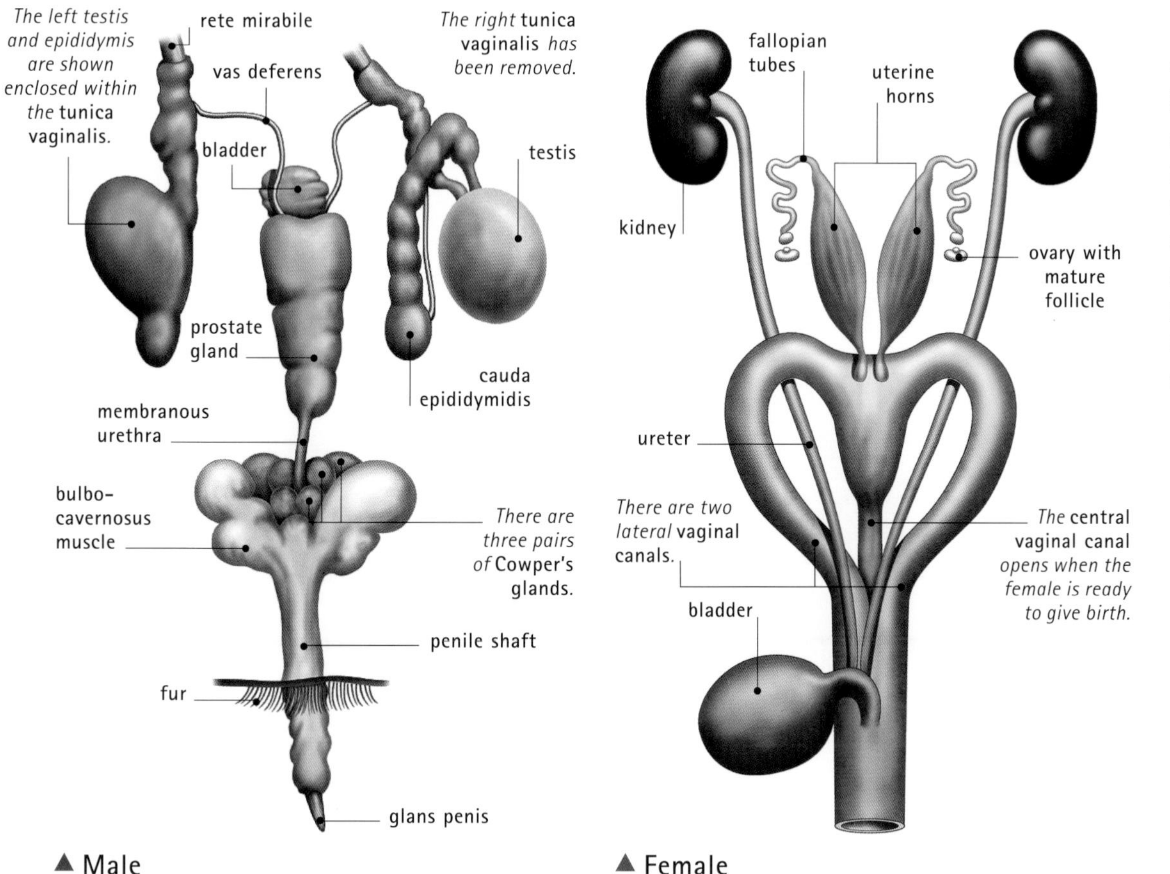

▲ Male

▲ Female

◄ **REPRODUCTIVE ORGANS**

Female kangaroos are not unusual among mammals in having two ovaries. However, uniquely, kangaroos also have two uterine horns, or wombs, and two vaginas for mating.

Female red kangaroos have four teats, but individual joeys use only one. The composition of milk provided by the mother kangaroo varies over time to meet the changing needs of her developing offspring. The milk gets thicker as the joey develops. By the time the joey leaves the pouch and begins exploring the outside world, it will be fueled by rich creamy milk that continues to flow from the same teat it has used all its life. Remarkably, a female nursing one infant who has left the pouch may already have another tiny baby inside the pouch, attached to a different teat and with its own supply of a quite different kind of milk. The milk provided for newborns is very thin and easy for the joey's underdeveloped gut to digest.

Red kangaroos have an amazing "production line" approach to reproduction. Because they live in a harsh environment where feeding conditions are unpredictable, and where droughts and other natural disasters such as bushfires are common, it pays to be able to breed whenever the opportunity arises—but also to be able to abandon the process at an early stage when the conditions are tough. Once pregnant, the females of placental mammals are committed to a long-term investment of their own bodily reserves. If food

◀ Newborn joey
Once inside the pouch, the tiny baby begins to suckle. It attaches to a nipple and remains there as it grows.

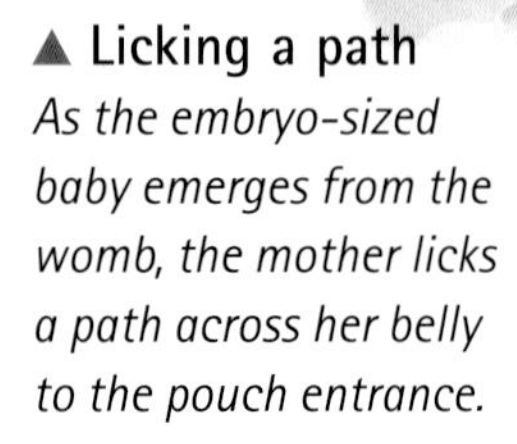

▲ Licking a path
As the embryo-sized baby emerges from the womb, the mother licks a path across her belly to the pouch entrance.

▲ In the pouch
After several months in the pouch, the young kangaroo starts exploring the world, returning to the pouch to suckle.

▶ One year old
The young kangaroo eventually leaves the pouch for good. It stays close to its mother while it learns to find its own food.

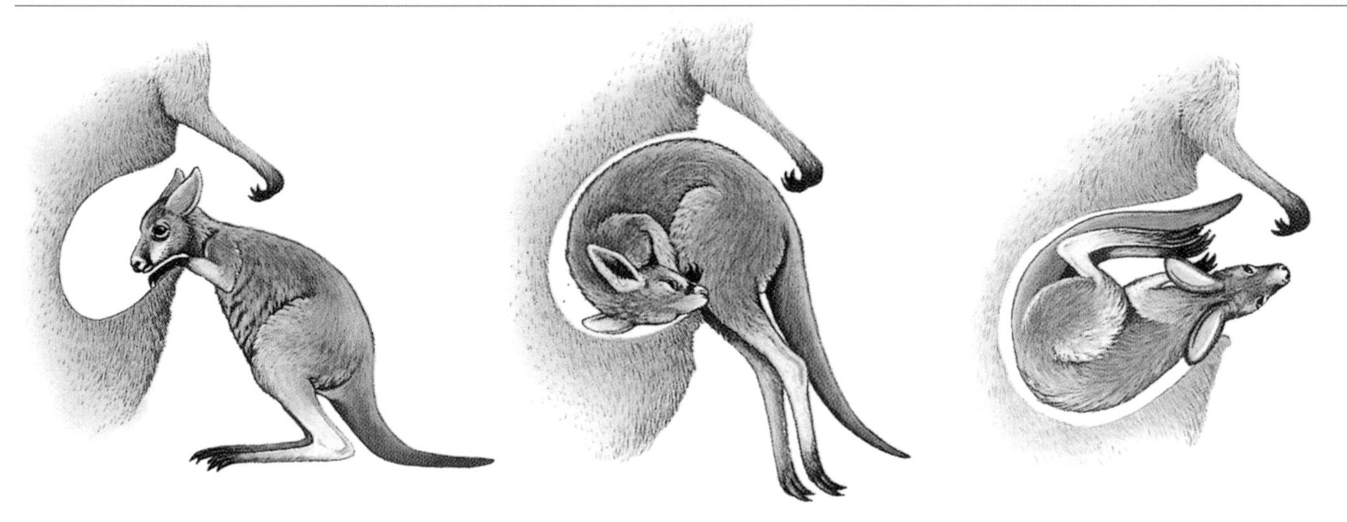

▲ **Joey acrobatics**
To get back into the pouch, the joey somersaults in head-first, then rotates its body so that its head can poke out.

◀ *The pouch is a comfortable place for the young joey—safe and with a ready supply of the right kind of milk.*

IN FOCUS

Incredible journey

A newborn kangaroo is less than 1 inch (2.5 cm) long. It has no eyes and no fur, and its back legs are mere buds. That it can survive at all outside the mother's body seems miraculous. Yet this tiny embryo manages to find its way from the birth canal all the way up the mother's belly, through the forest of fur and into the pouch. The only help the female gives her infant is to sit back on her tail to reduce the risk of the newborn's falling off her belly, and to lick a path through her belly fur. The blind newborn gropes its way along this saliva trail using a swimming motion to haul itself through the fur using just its forelimbs. The journey takes about three minutes, after which the tiny creature is completely exhausted. Once inside the pouch, it attaches itself to a teat, which becomes its life-support system for the next six months or more.

is in short supply when the baby is born, it may not survive, and the mother has reduced her own chances of survival for no reason. By giving birth to very tiny young after a short pregnancy, the female kangaroo makes a relatively small initial investment, which she will continue with the young in the pouch as long as conditions remain favorable.

The red kangaroo's estrus cycle is two days longer than its gestation period. Unlike other mammals, whose estrus cycle stops when they become pregnant, that of female kangaroos continues and a female will be able to mate successfully just a day or two after giving birth. The cycle then stops only if the first baby, now in the pouch, latches on to a teat and begins to suckle.

The same hormones that regulate the production of milk send the second baby—at this point an embryo containing no more than 90 cells—into a state of suspended animation known as embryonic diapause. The second embryo resumes development only when changes in milk production signal that the first offspring will soon be ready to vacate the pouch. Joey number two will be born within two days after its older sibling leaves the pouch for the last time, and the female will soon mate again. In this way, the female wastes no time between births; and if the joey in the pouch is lost, the female already has a replacement baby developing and need not wait to come into estrus or find a mate.

AMY-JANE BEER

FURTHER READING AND RESEARCH

Macdonald, David. 2006. *The Encyclopedia of Mammals.* Facts On File: New York.

Lion

ORDER: Carnivora FAMILY: Felidae SPECIES: *Panthera leo*

The second largest of the big cats, the lion was once widespread throughout Africa, the Middle East, and South Asia, as far as eastern India. Now it lives only in scattered areas of Africa south of the Sahara as far south as Botswana, and a small population inhabits the Gir Forest region of northwest India. The lion is powerfully built for hunting large, fast-running prey over open savanna.

Anatomy and taxonomy

All animals and other organisms are classified in groups based mainly on shared anatomical features. The features usually indicate that the members of a group have the same ancestry, so the classification shows how the organisms are related to each other, and to extinct fossil forms. Lions belong to the cat family, which is part of the mammalian order Carnivora, the carnivores.

- **Animals** All true animals are multicellular organisms that can move using muscles and have the ability to respond rapidly to stimuli. Animals obtain nutrients by eating other organisms, digesting their complex tissues to break them down into simpler molecules. Animals' bodies use these simpler molecules to provide energy or to build tissues.

- **Chordates** A chordate has a strong, flexible rod along its back, called a notochord. This supports its body and makes its muscles work more effectively. Most chordates retain the notochord throughout life, but some simple types such as sea squirts lose it as they mature.

- **Vertebrates** The notochord of a vertebrate forms the basis of a flexible backbone made up of units called vertebrae. The vertebrae and other skeletal units are made of bone or cartilage and provide anchorage for muscles that are mirrored on the left and right of the body. This arrangement is called bilateral symmetry. A vertebrate also has a brain enclosed within a cranium, or skull, and the group is sometimes called the Craniata.

- **Mammals** Mammals are warm-blooded vertebrates that feed their young on milk produced by the females. Typical mammals are insulated by a covering of fur or hair, which is unique to mammals. A mammal's jaw is hinged directly to its skull, unlike the jaws of all other vertebrates, and its red blood cells do not have nucleii.

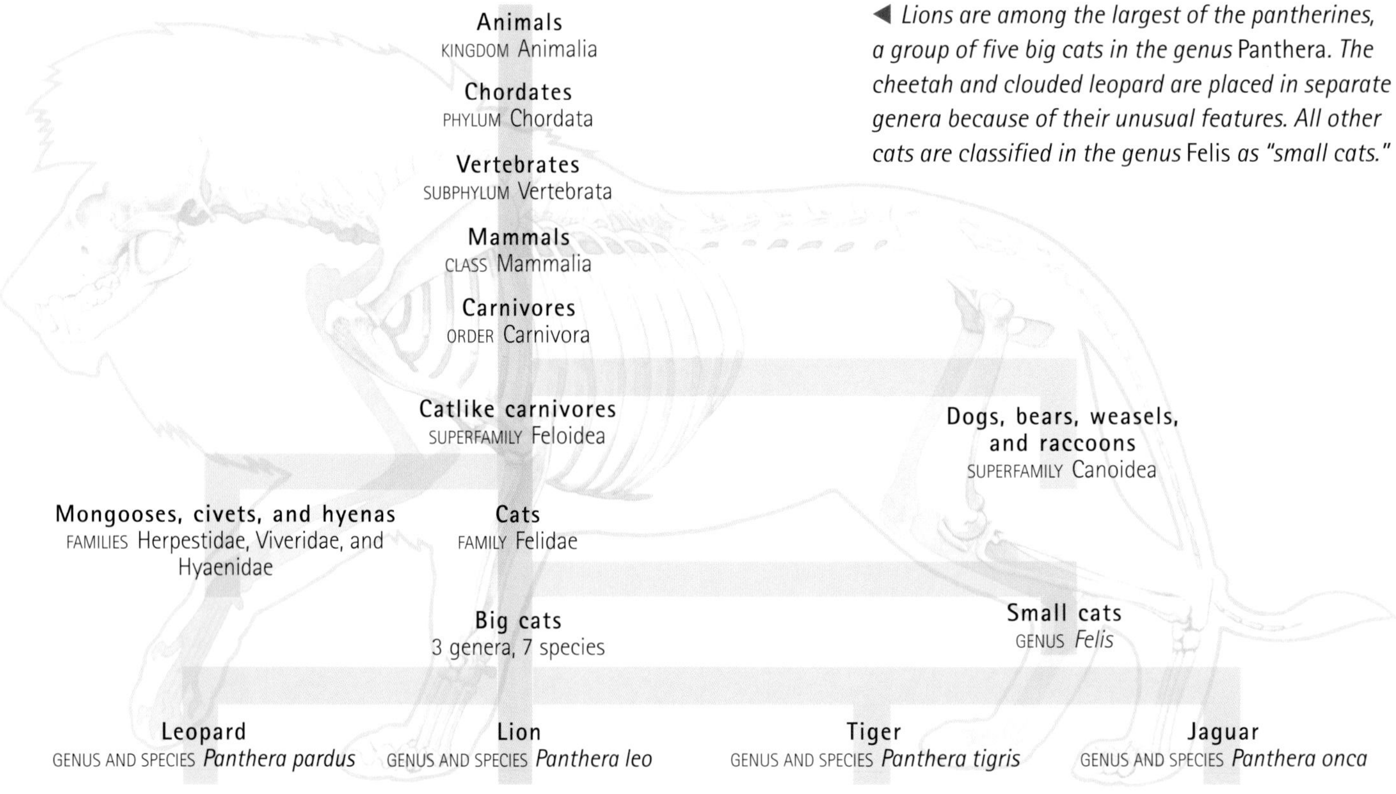

◀ *Lions are among the largest of the pantherines, a group of five big cats in the genus* Panthera. *The cheetah and clouded leopard are placed in separate genera because of their unusual features. All other cats are classified in the genus* Felis *as "small cats."*

• **Placental mammals** Placental mammals give birth to live young. Unlike the marsupials and monotremes, placental mammals nourish their unborn young during pregnancy with nutrients that pass from the mother's bloodstream to that of the embryo through an umbilical cord and placenta attached to the wall of the uterus.

• **Carnivores** Mammals of the order Carnivora are equipped with cheek teeth that are modified into shearing blades. These carnassial teeth have evolved for slicing through flesh, and most of the species in the order are meat eaters. They include the cats, dogs, hyenas, and weasels. However, the order also includes the omnivorous bears and raccoons, and the mainly plant-eating giant panda.

• **Cats** The cats are the most exclusively carnivorous of the Carnivora. They do not have chewing teeth, so they cannot deal with vegetable food. The muzzle is relatively short, enabling the jaw muscles to exert maximum force on the long, sharp canine teeth. This feature helps them kill prey efficiently. Cats' bodies are well suited for stalking and ambushing prey rather than for long pursuits, and their feet typically have sharp, retractable claws that they use as weapons.

• **Big cats** Although similar in form to the small cats (genus *Felis*), big cats are generally larger, more powerful, and distinguished by the ability to roar. They concentrate on killing large prey with a suffocating throat hold, rather than killing small animals with a quick neck bite in the manner of small cats. Most of the seven species of big cats belong to the genus *Panthera*. Exceptions are the cheetah, which is adapted for unusually high-speed pursuit, and the largely tree-living clouded leopard.

• **Lion** Second only in size to tigers among the big cats, lions are powerful enough to kill large, dangerous prey such as zebras and African buffalo. Unusually, lions live in groups and often hunt cooperatively, behaviors that may be linked to their open savanna habitat. Male lions look quite unlike females. Such sexual dimorphism is unique among the cats. There are two main forms of lions: the Asiatic lion *Panthera leo persica*, and the African lion *Panthera leo leo*, which is sometimes divided into four living subspecies.

▲ *Two lionesses visit a water hole for a welcome drink.*

FEATURED SYSTEMS

EXTERNAL ANATOMY These big, powerful cats display pronounced sexual differences; males are bigger than females and typically have a mane. *See pages 666–669.*

SKELETAL SYSTEM This is adapted for power and agility, with strong limbs and a flexible spine. The skull has a short muzzle that exerts maximum biting power. *See pages 670–671.*

MUSCULAR SYSTEM Lions are heavily muscled. This characteristic gives them extreme strength and the ability to kill animals several times their own weight. *See pages 672–673.*

NERVOUS SYSTEM Acute senses help locate and target prey, and intelligence enables tactical skill and social interaction. *See pages 674–676.*

CIRCULATORY AND RESPIRATORY SYSTEMS The deep chest contains large lungs for absorbing high volumes of oxygen during pursuit of prey. *See pages 677–678.*

DIGESTIVE AND EXCRETORY SYSTEMS A short digestive tract is specialized for processing meat. *See pages 679–680.*

REPRODUCTIVE SYSTEM Males select females in estrus and mate with them repeatedly, guarding them from rivals. *See pages 681–683.*

External anatomy

CONNECTIONS

COMPARE the powerful build of a lion with that of the lean, lightweight ***WOLF***. A wolf is capable of running long distances in pursuit of prey. In contrast, a lion is well equipped for ambushing prey with maximum impact.

COMPARE the male lion's mane with the antlers of a male ***RED DEER***. In both animals the adornments are used to intimidate rival males and possibly to impress females as well.

The lion is a big, powerful animal. A mature male typically weighs up to 530 pounds (240 kg), and the heaviest ever recorded weighed a colossal 690 pounds (312 kg): that is as much as four grown men. A male lion looks even bigger than it really is, thanks to the luxuriant mane that covers its head and shoulders. The mane is often darker than the short, sandy to reddish yellow fur that covers most of the body apart from its upturned tail tip. The tail tip has a dark tuft of fur, which is a feature unique to lions, and it conceals an equally unique horny spur that has no obvious function. The mane partly hides a pair of relatively short, mobile ears, which are black on the back.

*The **muzzle** is short and broad, so the jaw muscles exert maximum force on the daggerlike canine teeth that the lion uses to kill its prey.*

*Sensitive **whiskers** help the lion feel its way through undergrowth in the dark, when it does much of its hunting, and avoid making noises that could alert its prey.*

*The **mane** extends down the lion's back in some races. It grows longer, thicker, and darker as the animal gets older.*

*The **forelimbs** are powerfully developed to allow the lion to grapple with strong prey animals and pull them to the ground.*

▶ Male lion

Bigger and heavier than a lioness, a male lion looks even more imposing because of the thick, shaggy mane around the head, neck, and shoulders.

The mouth area, whiskers, and belly of a lion are whitish, and some individuals are virtually white all over. These white lions are not albinos, which lack color pigments throughout their body. White lions lack color pigment only in their skin and fur. Such white lions are very rare, partly because they are caused by a recessive gene that is usually masked by normal color genes, and partly because very pale lions are conspicuous and so find hunting difficult. They are more likely to go hungry and fail to breed. Very pale lion cubs also make easy targets for spotted hyenas and other enemies. At the opposite extreme, very dark lions are also rare, and totally black ones are unknown. However, some mature males have a blackish mane.

IN FOCUS

The lion's mane

The magnificent mane of a mature male lion can vary from a light, tawny yellow to dark brown or even black. It takes five to seven years to achieve its full length, and tends to get darker each year. Female lions may prefer males with a dark mane, but males use their mane primarily to impress rival males. The most extensive mane belonged to the Barbary lion, a North African subspecies that is now extinct, although the trait still shows up in some captive lions that may be descended from Barbary lions. By contrast the Asiatic lion has a relatively short, sparse mane, and some African males have no mane.

▶ **HEAD**
As well as increasing a male lion's bulk, the mane also helps protect the neck, which is an obvious target during fights.

The **hind limbs** *are strongly built, ideal for sprinting speed and leaping when the lion is attacking prey.*

The **tail tuft** *normally hides a horny spur, but some lions have just the tuft and no spur.*

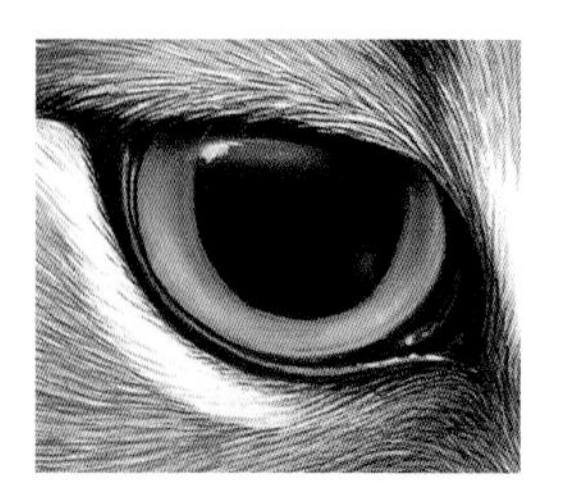

▲ Lion, night

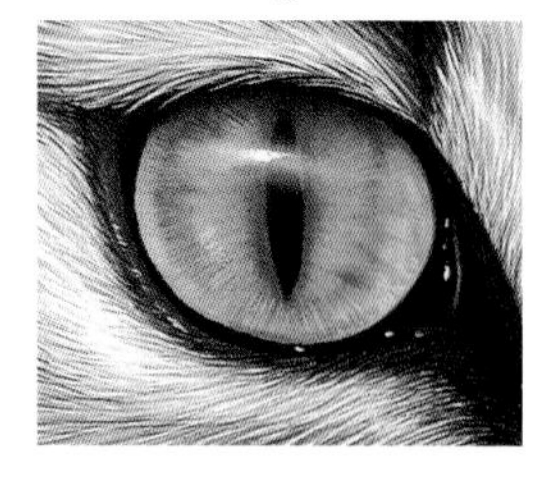

▲ Small cat, day

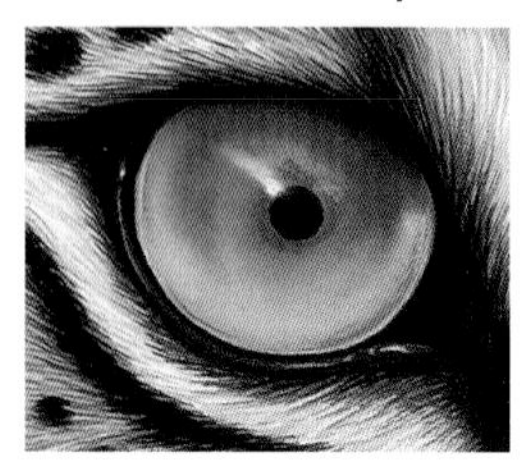

Jaguar, day

▲ EYES

The eyes of cats function well by day and by night. At night, a cat's pupils open wide, almost to the edge of the iris, to allow as much light as possible to enter the eye (lion, top). During the day, the pupils contract to prevent glare. Then, in the case of a small cat (center), the pupils become narrow slits, or in the case of a large cat such as a lion or jaguar (bottom), the pupils become small circles.

COMPARATIVE ANATOMY

Lions and tigers

Even bigger and more powerful than a lion, a tiger is in many ways its Asian equivalent. It, too, is an ambush predator, which uses its immense strength and sharp claws to pull down animals bigger than itself. However, unlike a lion it nearly always hunts alone, making the most of the cover available in its forest habitat. A tiger's striped coat, so striking when seen in isolation, provides perfect camouflage among the shadows and light of a tropical forest clearing, disguising the big cat's outline from its target until it gets close enough to strike.

▼ Tiger

The stripes of a tiger's fur break up its outline in the long grasses in which it hunts.

EVOLUTION

High-speed killers

The pantherine cats evolved as a result of climate change that occurred more than 5 million years ago. Earth became drier, leading to the expansion of tropical grasslands like the African savannas. Within 1 million years new types of grazing animals had evolved to live on the grasslands, including fast-running antelopes and gazelles. They were too fast for many hunters, including the saber-toothed cats, but a new lineage of faster cats appeared that could catch them. The faster cats included the direct ancestors of cheetahs, which evolved more than 2 million years ago. Eventually this line gave rise to the more powerful pantherines, and the first lions evolved about 600,000 years ago.

Heavyweight hunter

Everything about a male lion is massively built. It has a heavy head with a broad muzzle, powerful jaws, and very long canine teeth. Its body is deep-chested and sturdy, and its limbs are thick and muscular, with large paws. Each paw is equipped with strong claws that retract into sheaths to keep them sharp and maintain their effectiveness as weapons.

A female lion, or lioness, is similar to a male but is less massive, with no mane. The smaller size gives a lioness a much sleeker, but less imposing, appearance. Her smaller size limits the maximum size of prey she can bring down. A male lion can use his weight to topple a full-grown male giraffe weighing up to 1,100 pounds (500 kg), but his weight and bulk are a handicap in hunting faster, more agile prey such as antelopes and gazelles. Females can stalk these more effectively, being less visible, and they are quicker and more maneuverable

CLOSE-UP

Black for danger

Many wild cats have distinctive patches on the back of their ears. Tigers' ear patches are white, ringed with black, but lions' ear patches are black. The function of the patches is not known, but they probably help the animal communicate its mood. When a cat is angry it flattens its ears down against its head, and this shows off any markings. When lions are asserting their status or competing over a kill, they can avoid damaging fights by flashing their ear patches to show rivals they are dominant.

during the final pursuit. These attributes more than make up for their lack of sheer power, and females are generally able to kill a greater total weight of prey than males. When lions live together in a social group, or pride, the females do most of the hunting.

Camouflage

When a lion is stalking prey by daylight, its tawny yellow coat provides camouflage amid long, dry grasses on the African savanna. The camouflage allows the hunter to creep up close to its quarry before launching an explosive attack. This element of surprise is vital because the lion's powerful physique is ideal for ambush hunting rather than the long, potentially exhausting chases used by other carnivores such as wild dogs and wolves.

▼ *Although lions drink regularly when water is available, they are capable of obtaining their moisture requirements from prey and even plants (such as the tsama melon in the Kalahari desert), and can thus live in very arid environments. This drinking lion's mane shows it to be an adult male.*

Skeletal system

CONNECTIONS

COMPARE the short muzzle of a lion with the long muzzle of a ***GRIZZLY BEAR***, which has chewing molar teeth as well as shearing carnassial teeth. The combination of molars and carnassials enables the bear to eat a wide variety of foods.

Like all mammals, lions have a strong internal skeleton made of bone. The bone itself is formed from hard but brittle calcium phosphate crystals, embedded in a tough, flexible protein called collagen. The combination gives the bone strength and a certain amount of springiness, which is vital if it is not to break under stress.

Flexible spine

All the bones of the skeleton are ultimately attached to the central spine. This is built up from a long chain of separate vertebrae, linked together by elastic ligaments. Each vertebra fits tightly against its neighbor, but most are able to move in relation to each other. The 7 neck vertebrae are particularly mobile, allowing a lion to twist its head around to groom its fur, but the 13 thoracic and 7 lumbar vertebrae are also much more mobile than those of most animals. This gives a lion a very flexible spine, which is a feature common to all cats. When a lion runs, this flexibility extends its stride length by allowing it to open its stride and bring its hind legs farther forward. This characteristic is developed to a maximum in cheetahs, Earth's fastest-running animals.

A lion has broad shoulder blades, which provide attachment for its powerful forelimb muscles, but the collarbone that links the scapulae to the sternum in most animals is reduced to a thin sliver of bone attached by ligaments. This arrangement enables the shoulder blades to move more freely, increasing a lion's stride length and speed. A lion also walks and runs on its toes; this stance, which is typical of fast runners, is called digitigrade. Otherwise, its leg bones are not as specialized for running as those of a dog or a cheetah. The lion's forelimbs are especially heavy and very strong. This strength and weight are necessary to enable the lion to overpower large prey.

▼ **African lion**
The skeleton is similar in structure to that of other species of cats.

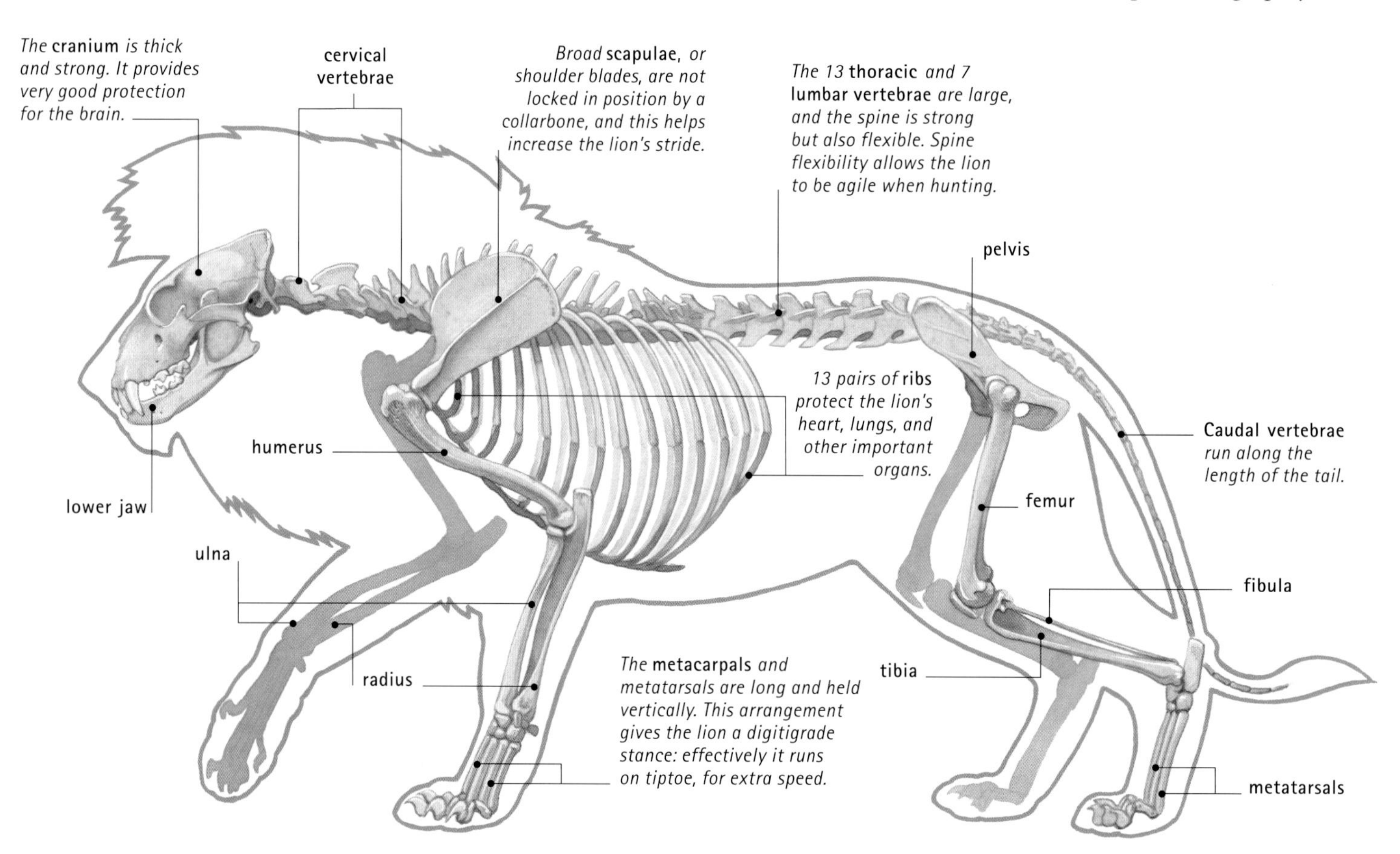

Massive skull

A lion's skull is strong and heavy, with deep ridges and hollows for the attachment of powerful jaw muscles. The leverage exerted by these muscles is increased because the muzzle is short relative to that of other carnivores, with the daggerlike canine teeth nearer the jaw hinge. The extra power, as well as the gap behind the canines, allows a lion to drive its teeth deep into its victim's throat and kill the prey more quickly than by suffocation alone.

The short muzzle, which is common to all cats, leaves less room for long rows of cheek teeth, so, unlike a dog, a lion has no flattened, chewing molars at the back of its mouth. The cheek teeth are relatively small, blade-shaped carnassials, which act against each other like scissor blades to shear through skin, sinew, and muscle as the lion devours its prey.

▼ SKULL

African lion

The short muzzle enables the jaw muscles to bite with great power, and the jaw hinge is strong enough to cope with the stresses of making a kill.

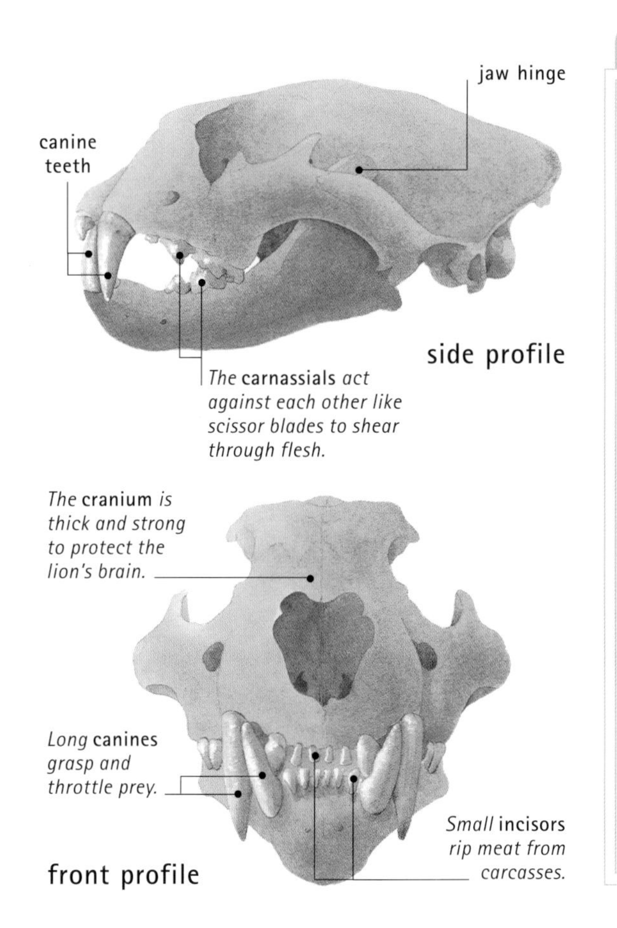

IN FOCUS

Extendable claws

The bones at the end of each toe are usually pulled up and back by strong elastic tendons, withdrawing a lion's claws into sheaths of skin. This arrangement keeps the claws from becoming blunted by contact with the ground as the lion walks or runs. When it needs them for seizing prey, the lion contracts muscles that straighten the toe bones and make the sharp claws protrude from their sheaths. Cats also need sharp claws for climbing. Less bulky leopards regularly haul their kill up into trees to keep it from being stolen by hyenas. Even heavyweight lions may take to the trees when seeking shade.

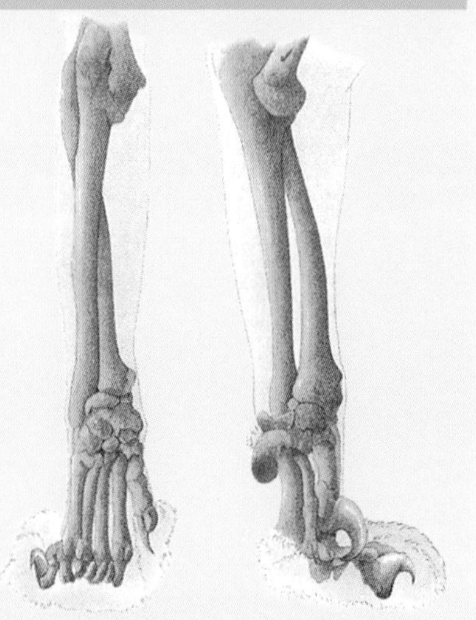

EXTENDABLE CLAWS

When the toe bones are pulled up, the lion's sharp claws are kept in their protective sheaths (left). When flexor muscles contract, the toes straighten and the claws are exposed.

IN FOCUS

A lion's roar

Only big cats can roar, and lions roar loudest of all. The structure of the voice box, or larynx, and the hyoid bones makes roaring possible. The larynx is very large, and one of the bones that supports it is replaced by an elastic ligament 6 inches (15 cm) long. The ligament can stretch to 8 inches (20 cm), creating a wide air passage. The extra width allows a lion to make an extra-loud roar, which registers up to 114 decibels.

▼ *A roaring lion can be heard up to 3 miles (5 km) away.*

Muscular system

> **CONNECTIONS**
>
> **COMPARE** the muscle sytem of a lion with that of a smaller cat such as a ***PUMA***. The lion's massive muscle system enables the animal to subdue prey larger than itself.

As you might expect, small cats catch small prey, and big cats such as lions catch bigger prey. Yet the increase in the size of their prey is out of proportion to their own size. While small cats concentrate on animals that are smaller than themselves, such as mice and rabbits, many big cats target prey that are substantially bigger than they are. Lions hunting alone regularly kill animals twice their own weight, and sometimes more. Achieving this kind of feat takes a lot of power, and lions are built accordingly.

Contracting fibers

A lion's skeletal muscles are the muscles attached to its bones that give it the power of movement. These muscles consist of long fibers built up from alternating filaments of the proteins myosin and actin. When a nerve impulse stimulates a muscle to contract, projections from the thick myosin filaments attach to the thinner actin filaments and haul them alongside, so the thin filaments slide between the thick ones. This shortens the muscle fibers and makes the entire muscle contract.

When the nerve impulse is switched off, the links between the filaments are released, allowing them to slide apart again and extend. This aspect of muscle action is relatively passive, so the skeletal muscles are arranged in pairs that work in opposition. The contraction of one muscle extends the other, and vice versa. One of the pair is often stronger than the other. For example, a lion has very powerful muscles for closing its jaws on its prey but only weak muscles to open its jaws for another bite.

Extreme force

Since the lion is an ambush hunter, its skeletal muscles are adapted for strength and explosive action rather than endurance. Its hind legs have massive thigh muscles, which give it the power

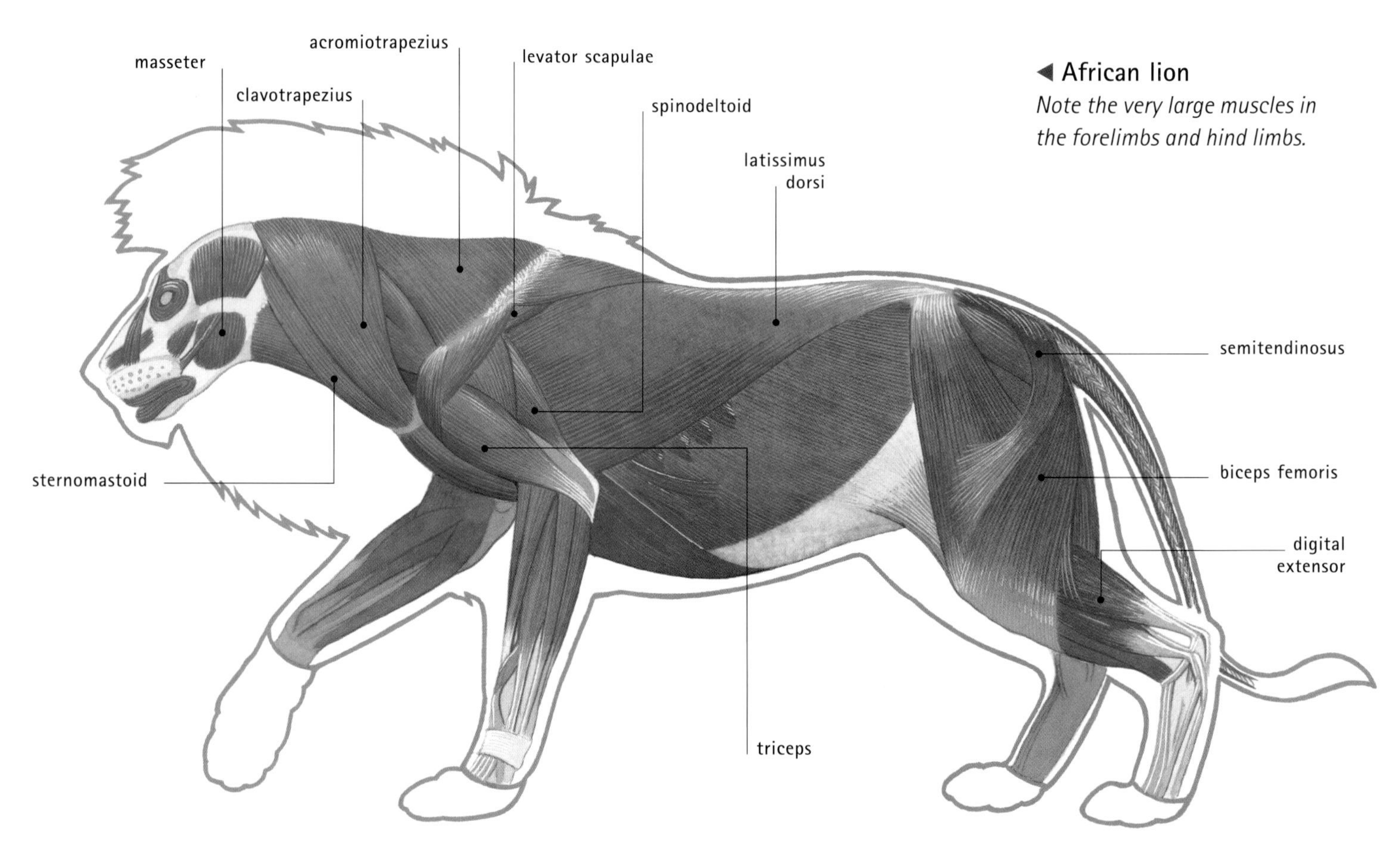

◀ **African lion**
Note the very large muscles in the forelimbs and hind limbs.

COMPARATIVE ANATOMY

Cheetah

Most big cats are heavyweight killers, but the cheetah is a lean lightweight with long, slender limbs tipped with blunt, doglike, only partly retractable claws. These features enable the cheetah to pursue prey at great speeds. A hunting cheetah can run faster than any other animal: up to 60 miles per hour (95 km/h). It cannot maintain this speed for long, however, and the average sprint lasts for less than a minute. So although the cheetah looks like a feline greyhound, its hunting strategy is more like that of a lion. It stalks its prey until it is within about 100 feet (30 m) and then makes an explosive attack. Since it is much less muscular than a lion, a cheetah cannot kill prey animals bigger than itself; it usually targets small gazelles.

Cheetah

The muscles of a cheetah's forelimbs are much smaller than those of a lion, but this is not a problem for a cheetah, since it subdues smaller prey than its larger relative. The muscles of a cheetah's hind limbs can drive the animal forward at great speed when it is hunting prey such as antelopes.

to attack from a standing start, crouched in ambush. In a few seconds a charging lion can accelerate to 35 miles per hour (57 km/h) and overtake startled prey before the prey can reach its own top speed. If the lion gets within striking range, its hind limbs can propel leaps of nearly 40 feet (12 m).

On contact, the lion's forelimbs come into play. Muscles in its feet contract to straighten its toes and project its long, sharp claws, which the lion digs into the hide of its target's rump, back, or shoulders. The lion then uses its weight and powerful forelimb muscles to wrestle its victim to the ground. It needs prodigious strength to achieve this with strong animals like zebras, and it must work fast. A struggling zebra could easily smash a lion's jaw with a well-aimed kick, so the lion must get it on the ground as quickly as possible and throttle it with a powerful neck bite.

Nervous system

CONNECTIONS

COMPARE a lion's vision with that of a ***CHIMPANZEE***. Both animals have binocular vision. A lion needs it to gauge distances when hunting, and a chimpanzee uses it to judge the distance between branches as it moves through the trees.

Some predators, including many small cats, hunt small animals that are difficult to find but relatively easy to catch and kill. Lions and other big cats hunt mainly big animals that are easy to find but difficult to catch and kill. This strategy makes different requirements of lions' nervous systems, and in particular their senses and mental skills.

Sense and sensitivity

A fox hunting mice hidden in long grass needs a sharp sense of smell to follow scent clues, acute hearing to pinpoint faint rustles and squeaks, binocular vision to judge distances accurately, and lightning reflexes.

A lion stalking wildebeest on the open savanna has to overcome a different set of problems. Locating the wildebeest is not

IN FOCUS

Stealth and touch

A hunting lion relies on stealth, and its sense of touch is vital. Its head bristles with sensitive whiskers that allow it to feel its way through the night and avoid rustling vegetation that could betray its presence. The whiskers can even detect disturbances in the air, providing the lion with a tactile picture of its immediate surroundings. This is particularly valuable at night, when a cat's pupils are wide open, because in that state its eyes cannot focus on objects that are right in front of its nose.

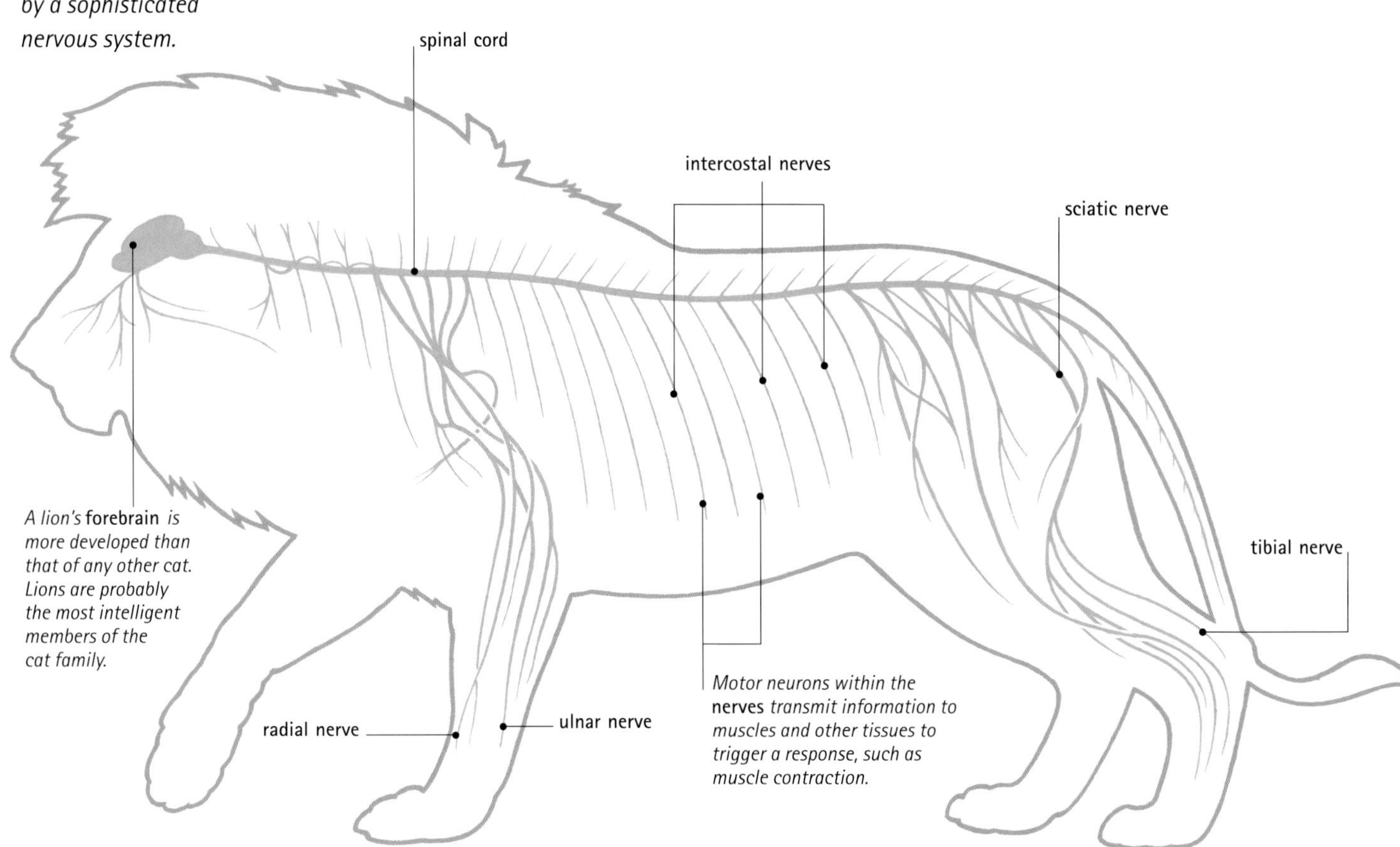

▼ **African lion**
Like all cats, the lion has keen senses and rapid reactions, all of which are coordinated by a sophisticated nervous system.

difficult, since prey are usually in plain view, grazing in great herds. So a lion does not need a refined ability to discriminate between scents that may lead it to prey. A lion's short muzzle contains a much smaller area of scent-receptive membrane than the long muzzle of a fox or dog, and the section of the brain that deals with scent processing is also relatively small. Although a lion uses its hearing a lot when hunting, its ears are not particularly large and sensitive compared with those of its smaller relative, the serval, which hunts hidden rodents in much the same way a fox does.

The importance of sight

Above all, a lion relies on vision, and like a fox it has sharp, binocular eyesight. Binocular vision relies on two eyes facing forward, each with a slightly different angle of view. When the two views are processed in the brain they create a three-dimensional image. This 3-D image is vital to an ambush predator such as a lion, which needs to know exactly how far it is from its victim before it makes its move.

A lion's eyes work well in dim light, thanks to a high proportion of supersensitive rod cells in the retina, compared with color-discriminating cone cells. There is also a mirrorlike membrane at the back of each eye that reflects light back into the retina to stimulate its cells as much as possible. These features of the eyes enable the lion to hunt confidently at night, when the big cat is less likely to be seen by its prey.

Intelligent cat

All of a lion's senses are linked to nerves that carry electrical nerve impulses to the spinal cord and brain. The sensory impulses often trigger rapid reflex reactions that do not involve the brain but instead send signals directly from the spinal cord to the muscles for immediate action.

As a stealth hunter, however, a lion often has to devise tactics for getting close to its prey. These may also involve other lions if they are hunting as a team. It must also learn to identify soft targets such as old or sick animals; a lion must learn to profit from its mistakes. So although instinct is important, a lion works more intelligently than many predators. It has a highly developed forebrain, the center of memory; and since it is the most social of cats, it is probably the most intelligent.

A lion's senses also keep it on its feet during twisting, turning, and leaping. In addition to sensory information from its ears and eyes, an

▲ The sense of sight is the most important for a lion in searching for prey, but a lion also has good senses of hearing and smell.

involuntary reflex helps a falling lion to right itself. In an automatic twisting reaction, the head rotates; then the spine and hindquarters align. At the same time, the lion arches its back to reduce the force of the impact when all four feet touch the ground.

Territorial animals

Scientists believe that (unlike other carnivores) lions and other big cats rely less on their sense of smell, or olfaction, to locate prey. However, smell seems to be important when big cats communicate with other members of their own species.

Lions of both sexes mark the boundaries of their pride territories with scent, using urine, feces, and scented secretions from glands between their toes. The urine of females in estrus also contains a distinctively scented chemical called a pheromone. Male lions can identify this using an organ in the roof of the mouth called the vomeronasal or Jacobson's organ. When a male detects the scent he pulls his upper lip back with his teeth bared in a grimace called the Flehmen response. The response helps the sensory cells of the vomeronasal organ analyze the pheromone and assess the female's breeding condition.

CLOSE-UP

Retina cells

The mirrorlike membrane at the back of a lion's eyes is called the tapetum lucidum. It is formed from specialized platelike cells that intercept any light that passes between the cells of the retina and reflect it back to the highly sensitive rod cells. The eyes of many night hunters have this type of structure, but it is particularly well developed in lions and other cats, which have as many as 15 layers of reflective cells. They reflect up to 90 percent of the light that enters the eye, causing the "eyeshine" of a cat illuminated by flashlight.

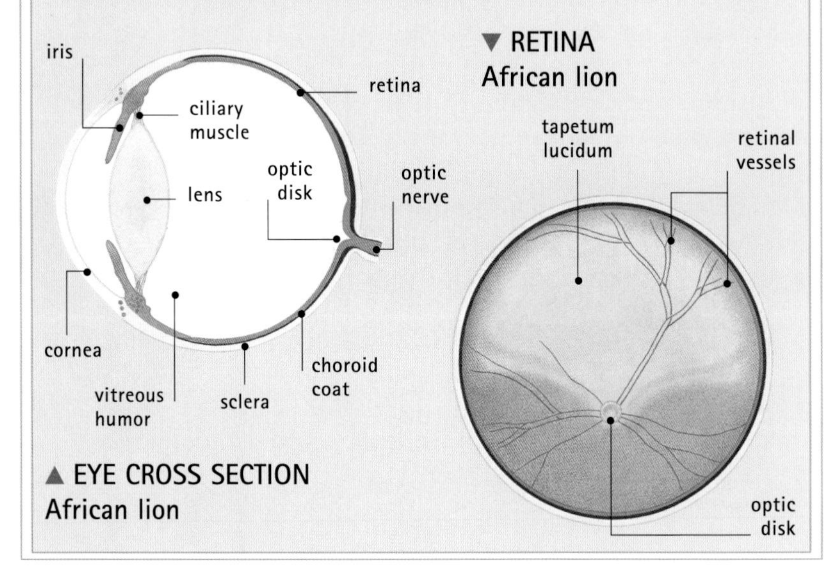

▲ EYE CROSS SECTION
African lion

▼ RETINA
African lion

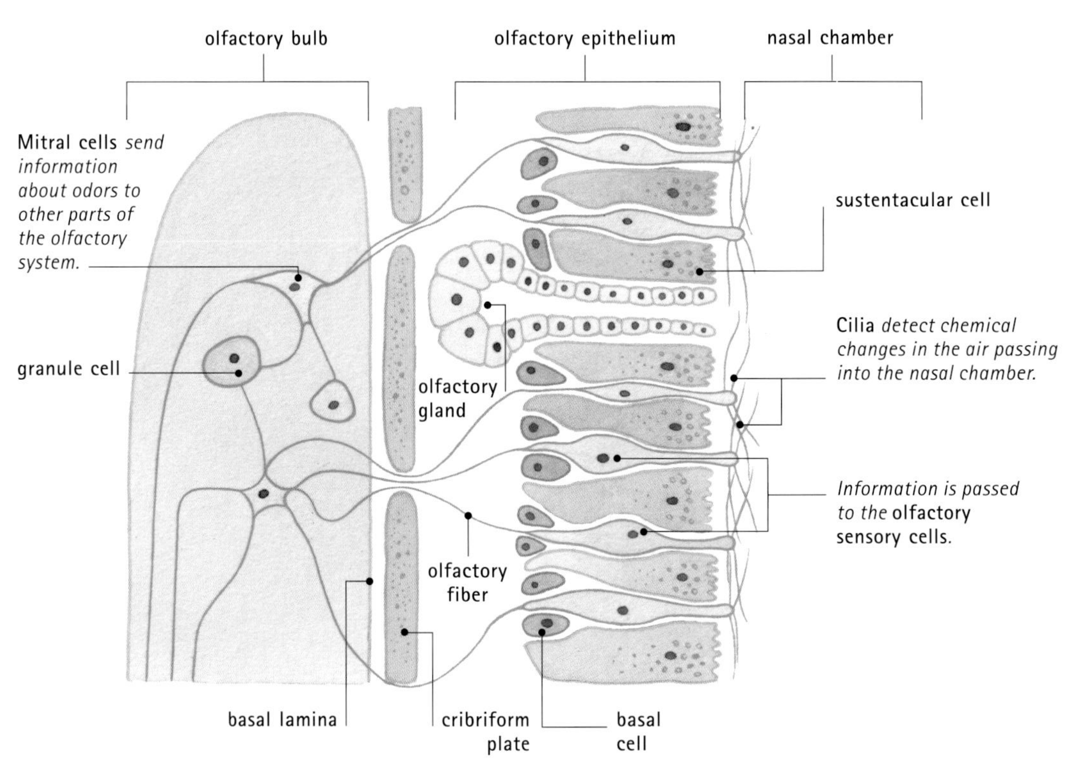

▶ OLFACTORY SYSTEM
Mammal
Lions and other mammals detect odors with their olfactory system. Cilia extend into the nasal chamber from each olfactory sensory cell. The cilia detect chemical changes in the air that the animal breathes. Nerve fibers extend from each olfactory cell to mitral cells. Fibers of the mitral cells together make up the olfactory tract, which goes to the brain.

Circulatory and respiratory systems

Hunting large prey is strenuous work. When a lion pursues and struggles with its prey, its muscles are using a lot of energy. Ultimately this energy is derived from both its food and the air that it breathes.

Digested food is absorbed into the blood, where some of its ingredients are turned into a simple sugar called glucose. The bloodstream delivers this blood sugar to the lion's muscles and other organs, along with oxygen taken up via its lungs.

When it reaches the cells of the muscles and organs, the sugar is mixed with the oxygen to trigger a chemical reaction called oxidation. It is virtually the same as burning, and like burning it produces energy. The energy powers the muscles, and the oxidized sugar is turned into carbon dioxide and water. The whole process is known as aerobic respiration.

Vital oxygen

If the lion is able to catch food, it usually has a good supply of blood sugar. Getting enough oxygen to oxidize the sugar is more difficult, so the lion has a very deep chest containing large lungs. Each lung is essentially a mass of small bubble-like sacs called alveoli, which are linked to a network of air tubes or bronchioles. These branch from larger tubes called bronchi, which are connected to the lion's windpipe, or trachea.

When the lion breathes in, it contracts the muscular diaphragm at the bottom of its rib cage. This makes its sealed lung cavity bigger, and expands its lungs so that they draw in air. The air passes into the alveoli, where oxygen passes through their thin walls and into a surrounding network of fine blood capillaries. At the same time, waste carbon dioxide and

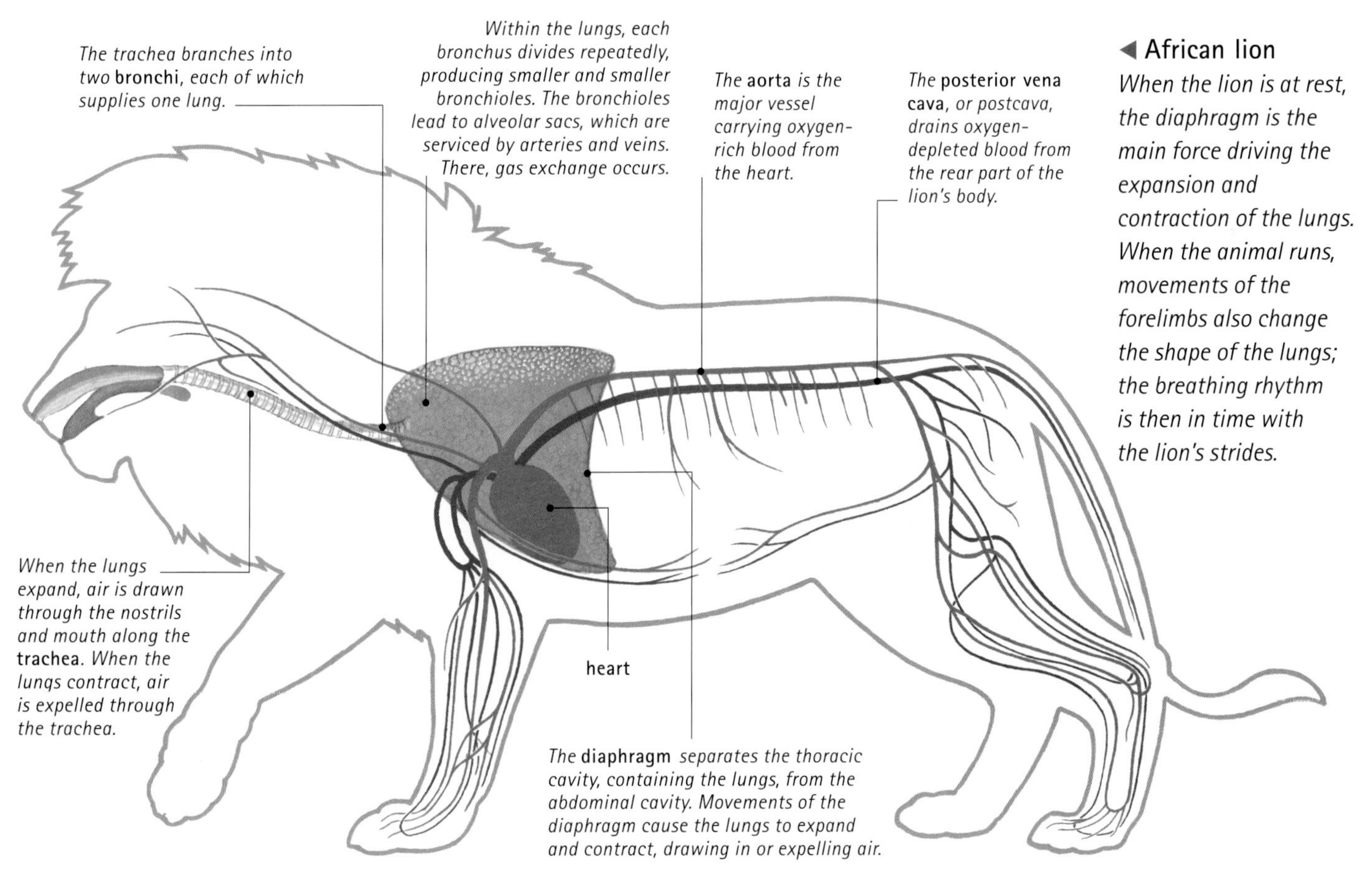

◀ **African lion**
When the lion is at rest, the diaphragm is the main force driving the expansion and contraction of the lungs. When the animal runs, movements of the forelimbs also change the shape of the lungs; the breathing rhythm is then in time with the lion's strides.

▶ *This male lion has just attacked and killed a zebra. When a lion mounts an explosive attack, it cannot deliver enough oxygen to its muscles. The lion then relies heavily on a process called anaerobic respiration.*

COMPARATIVE ANATOMY

Heavy breathers

When a lion catches its prey, it has to seize the prey with its teeth while breathing heavily through its nose to gather vital oxygen. Doing both at once is not easy, but the problem is far worse for the cheetah. At the end of its record-breaking sprint a cheetah needs much more oxygen to clear the buildup of lactic acid in its muscles. To make this possible, its nasal passages are proportionately larger than those of a lion. This leaves less room for the roots of its upper canine teeth, so its canines are proportionately shorter, limiting the size of prey that it can kill.

water pass out of the blood and into the air in the alveoli. When the lion relaxes its diaphragm, its lungs contract again, forcing the waste air out of its trachea.

Clogging the system

The lion's blood is pumped through its lungs by the right-hand side of its heart. Newly oxygenated blood from the lungs returns to the left-hand side of the heart, which then pumps it to the muscles and other tissues. These use the oxygen and replace it with carbon dioxide and water. The blood then returns to the right-hand side of the heart, which pumps it back to the lungs for more oxygen. When a lion is working hard, its heart pumps rapidly and it breathes very deeply. Yet it still cannot gather and deliver enough oxygen to its muscles. So when a lion mounts explosive attacks, it relies heavily on another process, called anaerobic respiration. This process releases energy without any immediate need for oxygen, by converting a sugary substance called glycogen into lactic acid. The lion can do this for only a short while, however; and if it sprints for more than about 1,000 feet (300 m), the lactic acid clogs its system and it must stop. The acid must then be cleared by using a lot of oxygen to change it to carbon dioxide, forcing the exhausted animal to breathe very heavily to recover.

IN FOCUS

Lion's blood

Mammalian blood is colored by red blood cells, which contain a red pigment called hemoglobin. Hemoglobin has a strong affinity for oxygen and allows blood to carry far more oxygen than it would otherwise. Each hemoglobin molecule binds to four oxygen molecules to form oxyhemoglobin. When blood is delivered to oxygen-depleted tissues, the oxyhemoglobin breaks up to release oxygen.

Digestive and excretory systems

Like all cats, lions are adapted for eating meat and nothing else. Most mammal carnivores—including bears, raccoons, dogs, and even weasels—eat vegetable foods regularly or occasionally and are equipped to chew and digest them. A lion may eat a little fruit to obtain water, but otherwise eats only flesh.

A lion has no chewing molar teeth, and its lower jaw can move only up and down, not from side to side in a chewing action. A lion can crush bones, although not with the same efficiency as a dog or hyena. A lion's lack of chewing ability is partly offset by sharp spikelets on its tongue that can shred meat and rasp it from the bone, but it swallows most of its food in big chunks sheared from the carcass by its scissorlike carnassial teeth.

A lion may also swallow a lot of meat very quickly because, like most hunters, it never knows when it will get its next meal. An adult male can eat up to 95 pounds (43 kg) at a sitting.

Meat is relatively easy to digest, so lions—and other cats—do not need a complex digestive system. A lion's stomach can hold a large amount of meat, but the animal's intestine is relatively short. Lions spend most of their time resting between infrequent hunting forays, and during these inactive periods the meat that they have eaten passes through the intestine, where it is digested.

Digestive enzymes

Enzymes in the digestive juices break the bonds that bind complex protein molecules together, reducing them to the simpler molecules of amino acids. Since amino acids are the building blocks of all proteins, the lion's cells can use them to make the proteins that its body tissues need. The amino acids are absorbed through the wall of the intestine into the bloodstream, along with some proteins and the digestion products of fats and glycogen.

CONNECTIONS

COMPARE the simple digestive system of the lion with the complex, multistage digestive system of a ***GIRAFFE*** or ***WILDEBEEST***. Both these animals are ruminants that eat leaves or grass, which are much more difficult to digest than meat.

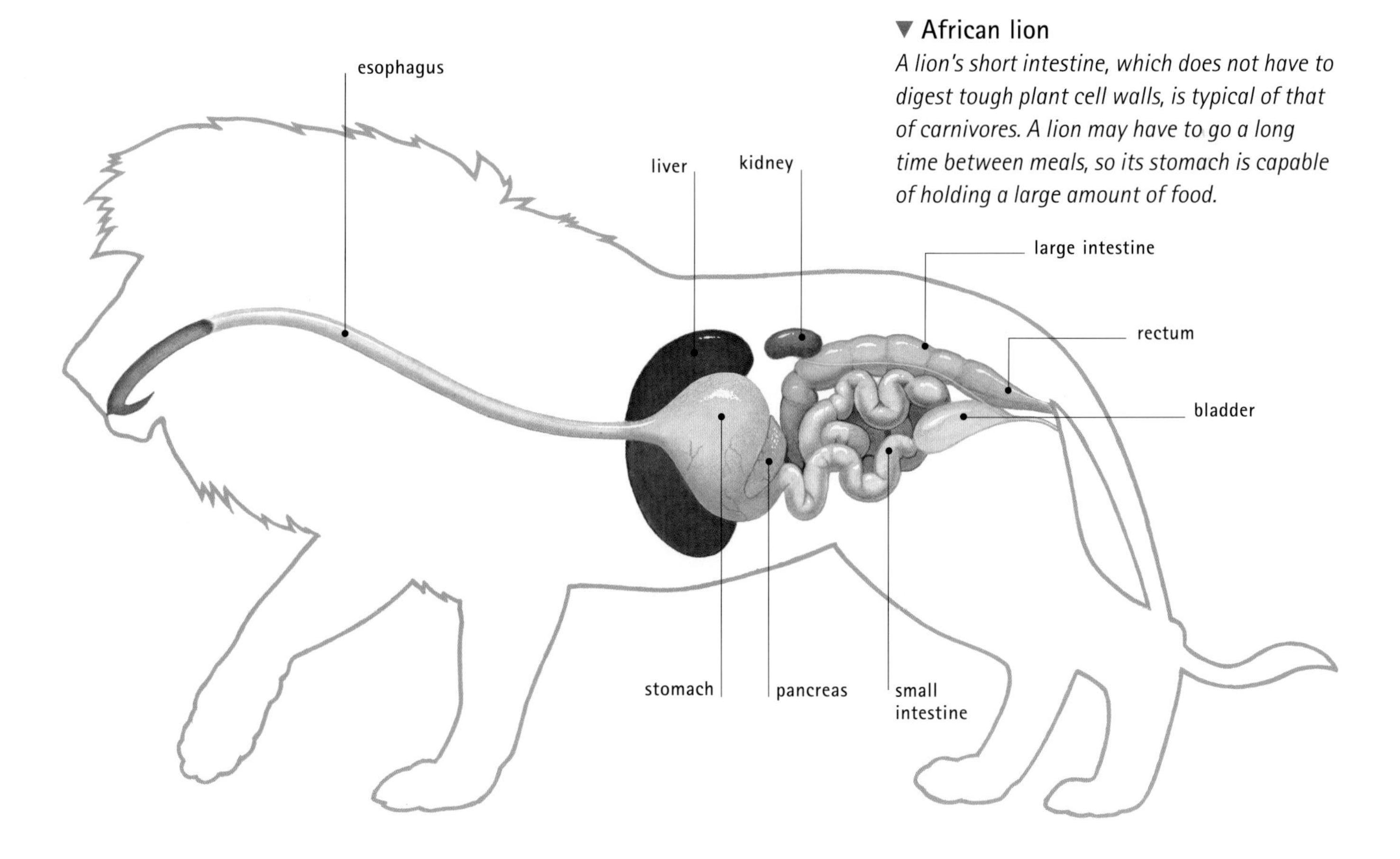

▼ African lion

A lion's short intestine, which does not have to digest tough plant cell walls, is typical of that of carnivores. A lion may have to go a long time between meals, so its stomach is capable of holding a large amount of food.

IN FOCUS

Glycogen store

One of the liver's functions is to turn carbohydrate foods into products that can be used by the lion's body. These products include the blood sugar glucose, which can be oxidized to provide instant energy for most of the lion's body processes. The liver also stores sugar in the form of glycogen, a substance that can be turned into lactic acid to liberate energy. By storing energy in this way, the liver ensures that there is no temporary shortage of blood sugar, and the glycogen also provides the fuel that powers the lion's charging attacks.

Since it eats no plant material, a lion does not need to digest complex plant carbohydrates such as starch or cellulose.

Excretion

All the blood flowing from the lion's intestines passes through its liver, which continues the work of processing the products of digestion and turning them into a form that the lion's body cells can use. The conversion process creates waste products, including some toxic substances. The liver cells are able to neutralize these, along with any other poisons in the blood, and transform them into a harmless nitrogen-rich substances that are carried in the bloodstream to the lion's kidneys. There they are filtered out of the blood, along with some water, and excreted as urine.

PREDATOR AND PREY

Team effort

Unlike other big cats, lions, and especially lionesses, often hunt as a team. This strategy works well in their savanna habitat: it is harder to mount a successful ambush in open grassland than in a forest with deep shadows and plenty of cover, so it helps if two or three lions surround the victim to cut off its retreat. When it comes to the actual kill, a single lioness is usually strong enough to do the job herself.

When lions have eaten their fill after a successful hunt, they may not need to eat again for two days or more. Prey animals are probably aware of this, and when the lions are clearly not hungry, antelope, gazelles, and zebras often graze close by, knowing that they are in very little danger of being attacked. Such behavior may provide a defense against other predators that are wary of encroaching on a lion's territory.

▼ *A group, or pride, of hunting lions is much more likely to make a successful kill on the savanna than a single lion.*

Reproductive system

Male and female cats of most species are very similar. Males are typically bigger and stronger, but they look much like the females. Male lions, by contrast, are quite distinctive, with a spectacular, shaggy mane. The mane reflects the fact that mature males have a specific role in lion society.

While female lions, like most cats, are superbly adapted for hunting, male lions are better equipped for fighting. Their enemies are other males that try to take over their territory and family group, or pride. Over many thousands of years of evolution male lions have become well equipped to fight, with extra muscle and a big mane that protects the vulnerable throat and makes them look more imposing without adding much extra weight. If the head male lion in a pride (the pride male) looks sufficiently impressive, he may be able to warn off a trespasser without a fight. Although pride males are aggressive toward trespassing males, they live happily alongside other mature males within the pride. The males are often brothers, but not always. At some point they will have succeeded in their own joint takeover bid, and moved in on a group of females that are nearly always sisters. The males' dominance over the pride typically lasts just two or three years before they are overthrown and evicted, or even killed by another coalition of males, but the females always stick together on their territory. This pride system is unique among wild cats, most of which live alone or with only their young.

Painful mating

All the mature females in a pride usually come into heat, or estrus, at the same time. The estrus period lasts for three to five days, and during this time each female may be courted by several pride males that can detect her condition by her scent. A male follows the

CONNECTIONS

COMPARE the placental development of an unborn lion with the development of a newborn ***KANGAROO***. The kangaroo does not have a placenta, and a baby gets all the nutrients it needs from its mother's milk while it develops in her pouch.

▼ MALE UROGENITAL SYSTEM
African lion

The male's penis is covered with tiny barbs, which are thought to stimulate ovulation in the female.

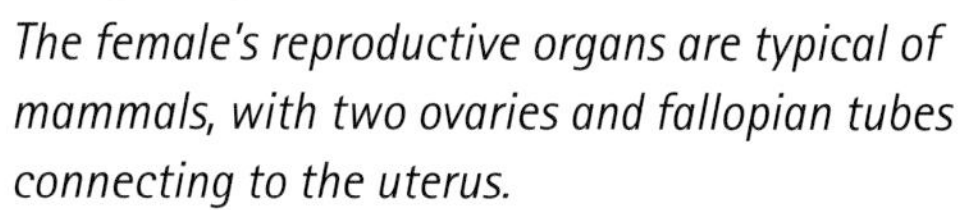

▼ FEMALE UROGENITAL SYSTEM
African lion

The female's reproductive organs are typical of mammals, with two ovaries and fallopian tubes connecting to the uterus.

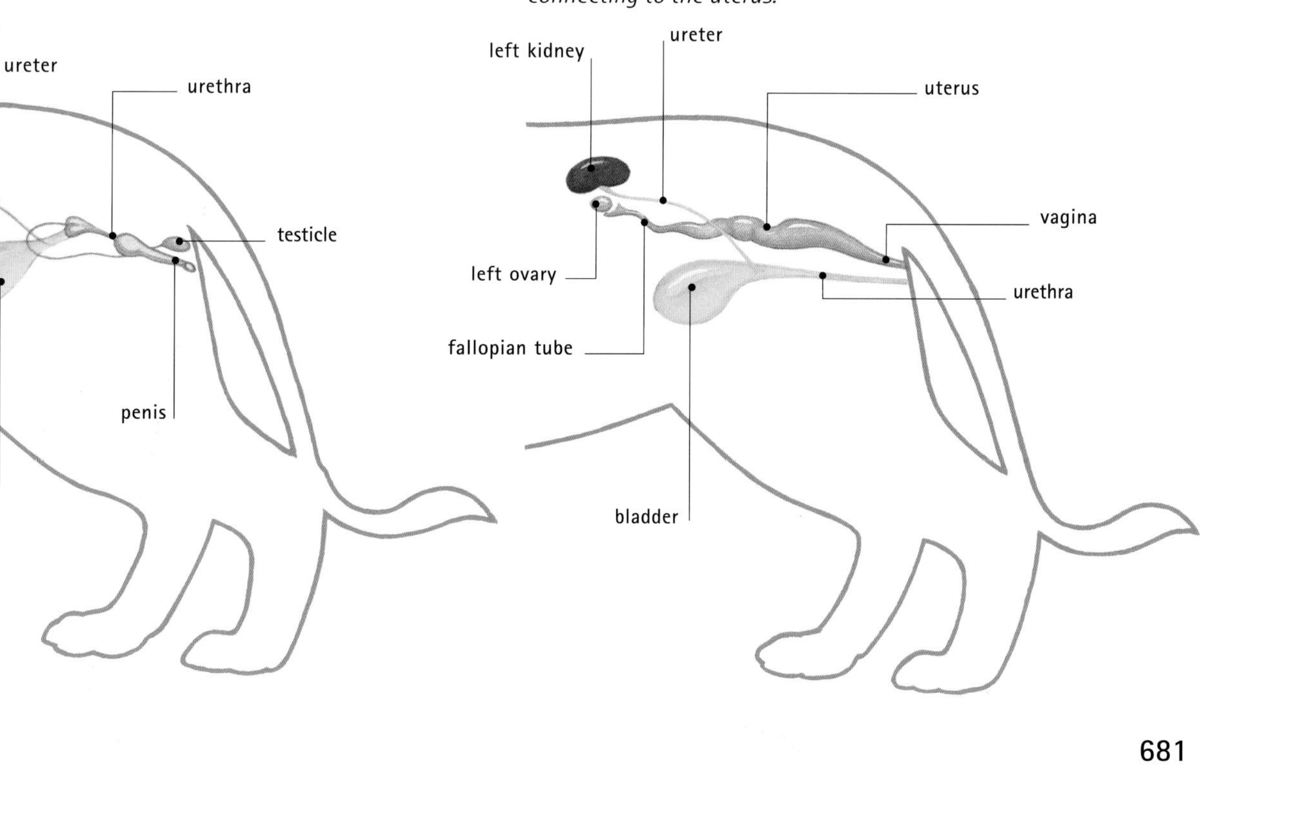

female around, and if he is persistent enough the female will lie down on her belly and allow him to mate with her. Since all the females in a pride are often in heat at once, each typically mates with just one pride male that guards her while mating every 20 minutes or so.

The male's penis is covered with tiny backward-facing spines that probably make his withdrawal painful for the female, and after dismounting he leaps aside quickly to avoid her teeth and bared claws. The function of the spines is uncertain, but during early matings they may stimulate the female's ovaries into releasing eggs, which are then fertilized by the male's sperm after one of the later matings.

Once a single sperm has penetrated an egg, the fertilized egg becomes implanted into the thick wall of the female's uterus, or womb, and

GENETICS

Infanticide

Since pride males are usually related to each other, they are also related to most of the cubs born within the pride. This makes them attentive uncles as well as fathers. When other males take over a pride, however, they know that any cubs already born are not related to them, and do not share any of their genes. Their aim is to change this as soon as possible, so they kill any young cubs they find. This has the effect of making their mothers come into estrus much earlier than usual, so the males get to mate sooner and father cubs that carry their own genes.

▼ *This pair of African lions are mating. If the female (below) is fertilized, she will undergo a gestation period of just under four months before giving birth. Litters usually contain two, three, or four cubs.*

EVOLUTION

Spotty cubs

When lion cubs are born their coat is spotted with dark brown rather than plain tawny yellow. The spots usually fade at the age of three months or so, but some lions stay faintly spotted throughout their life. The spots may help conceal the helpless cubs from marauding hyenas and eagles, but the plain adults seem equally well camouflaged on the dry, dusty savanna. Some zoologists believe that the spots survive from a distant time when lions of all ages were spotted like leopards; future evolution may discard the spots completely.

▼ *This lion cub is only a few weeks old. Very young lions have dark-spotted fur, but the spots usually fade long before maturity.*

starts to develop into an embryo. Each of two to four unborn young are connected to the mother by an umbilical cord. The cord is attached to a placenta, a temporary organ that allows nutrients to pass from mother to young.

Helpless cubs

The cubs are born after a short pregnancy of just 110 days. They are very small at birth, having only 1 percent of their adult weight, and they are born blind, deaf, and virtually helpless. The mother feeds them milk from her four nipples, and pride females that produce cubs at the same time may suckle each other's cubs. By 10 days old the cubs are fully active, and at 2 to 3 months they are following their mother to kills to eat meat. Despite this, they are not fully weaned until they are 5 or 6 months old. Young females generally stay and breed within their parents' pride, but sometimes groups of sisters leave to establish a new pride, either with or without accompanying males. Young males always leave the pride when they reach sexual maturity at around three years old. Brother lions often team up with unrelated males to form "bachelor gangs," and if they are lucky they will take over a pride and have an opportunity to father their own cubs.

JOHN WOODWARD

FURTHER READING AND RESEARCH

Kitchener, A. 1991. *The Natural History of the Wild Cats*. Natural History of Mammals Series. Cornell University Press: Ithaca, NY.

Macdonald, David. 2006. *The Encyclopedia of Mammals*. Facts On File: New York.

Sunquist, M. and F. Sunquist. 2002. *Wild Cats of the World*. University of Chicago Press: Chicago.

Lobster

PHYLUM: Arthropoda CLASS: Malacostraca
ORDER: Decapoda FAMILY: Nephropidae

Lobsters do not form a single taxonomic group but are found in some 17 families within two large groups called superfamilies. Although a few species live in deeper waters, most lobsters live close to coasts.

Anatomy and taxonomy

Scientists categorize all organisms into taxonomic groups based partly on anatomical features. The American lobster is one of dozens of species of marine lobsters in the groups Nephropoidea and Palinuroidea, which are within the order Decapoda (10-legged crustaceans).

• **Animals** Lobsters, like all animals, are multicellular lifeforms that gain their food supplies by eating other organisms. Members of the animal kingdom differ from most other multicellular organisms in their ability to move from place to place (in most cases, using muscles). They generally react rapidly to touch, light, and other stimuli.

• **Arthropods** Arthropoda is the largest and most successful phylum of organisms on Earth, yet despite its vast diversity the general body plan of arthropods is relatively constant. Arthropods have segmented bodies, although segments are typically fused in areas to form body regions: the head, thorax, and abdomen. Arthropods have a tough outer "skin" called an exoskeleton. It protects the internal organs and serves as an attachment point for muscles. To grow, an arthropod must shed, or molt, its exoskeleton. Arthropods have pairs of jointed appendages, such as legs, mouthparts, and antennae. Internally, all arthropods have a ventral (running along the underside) nerve cord and a dorsal (running near the back) blood vessel. The rear part of the dorsal vessel pumps blood, called hemolymph, around the body cavity (hemocoel).

• **Crustaceans** Most of the 55,000 or so species of crustaceans are aquatic. While many species live in the sea, some live in freshwater and a few live on land. Biologists classify crustaceans within the phylum Arthropoda along with insects, arachnids, and myriapods. Crustaceans have an exoskeleton, or outer skeleton, made of chitin and calcium carbonate. The crustacean body, like that of an insect, has three major sections: the head, thorax, and abdomen. Crustaceans are the only arthropods with two pairs of antennae—they are the first of five pairs of appendages on the head. The third pair of appendages form the mandibles or main jaws, and the fourth and fifth pairs are modified into the maxillae, or accessory jaws. Behind the head, the

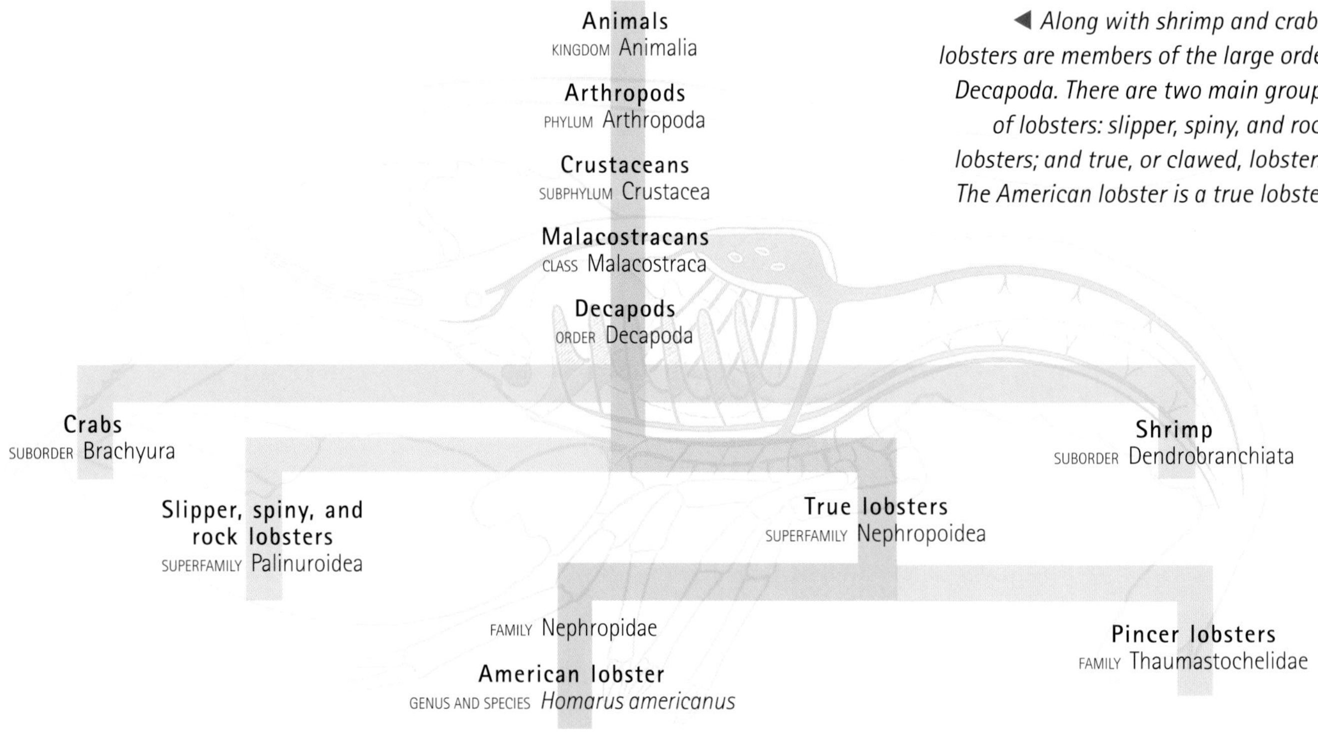

Animals
KINGDOM Animalia

Arthropods
PHYLUM Arthropoda

Crustaceans
SUBPHYLUM Crustacea

Malacostracans
CLASS Malacostraca

Decapods
ORDER Decapoda

Crabs
SUBORDER Brachyura

Shrimp
SUBORDER Dendrobranchiata

Slipper, spiny, and rock lobsters
SUPERFAMILY Palinuroidea

True lobsters
SUPERFAMILY Nephropoidea

FAMILY Nephropidae

American lobster
GENUS AND SPECIES *Homarus americanus*

Pincer lobsters
FAMILY Thaumastochelidae

◀ *Along with shrimp and crabs, lobsters are members of the large order Decapoda. There are two main groups of lobsters: slipper, spiny, and rock lobsters; and true, or clawed, lobsters. The American lobster is a true lobster.*

body is divided into segments (somites). Each segment typically bears a pair of appendages, and at least some of the appendages are biramous (two-branched).

• **Malacostracans** The class Malacostraca is the largest class of crustaceans, with more than 20,000 species. This class is extremely varied and includes at least 14 orders encompassing land-living pill bugs, ocean-dwelling krill, beach fleas, crabs, shrimp, lobsters, and freshwater sow bugs. The body of malacostracans typically has eight segments in the thorax and six segments in the abdomen. Each segment bears a pair of appendages. In most malacostracans, the stomach has two chambers and is followed by a midgut that has pockets called digestive glands where digestive enzymes are secreted and the products of digestion are absorbed.

• **Decapods** Decapods, or members of the order Decapoda (from the Greek *deka*, "ten"; and *podos*, "foot"), include lobsters, crayfish, shrimp, and crabs. As their name suggests, they have 10 legs—the last 5 pairs of appendages on the thorax. Decapods differ from other malacostracans in that the first three pairs of thoracic appendages (those directly in front of the legs) are maxillipeds, or feeding appendages. The cephalothorax (fused head and thorax) is often enclosed by a hard carapace (shell), which also encloses the gills. There are about 10,000 species of decapods, amounting to about one-fifth of all crustaceans. Most are marine, but crayfish and a few shrimp and crab species thrive in freshwater, and there are a few terrestrial species, such as the tropical land crabs.

• **Lobsters** There are two main groups of lobsters: the clawed lobsters (Nephropoidea) and the slipper, spiny, and rock lobsters (Palinuroidea). The clawed lobsters include the American lobster, which—like other clawed lobsters—has a forward-pointing spine on the carapace; in contrast, Palinuroidea species lack a permanent carapace spine. In all lobsters, the carapace is tubelike and the body as a whole is cylindrical, with a large abdomen. The first pair of walking legs are usually shaped into pincers, and the lobster crawls forward using its rear thoracic appendages as walking legs. The abdomen forms a broad "tail," which enables the lobster to move rapidly backward as an escape response. American lobsters spend the warmer months in shallow waters to depths of about 180 feet (55 m), close to the cool Atlantic coast of Canada and the United States. In winter, they live in deeper water.

▲ *Lobsters are large, sea-dwelling crustaceans. All species have numerous legs and other appendages, but not all have claws.*

FEATURED SYSTEMS

EXTERNAL ANATOMY Lobsters, like other crustaceans, are enclosed in a hard cuticle. This is shed periodically and replaced to allow the animal to grow. Jointed appendages growing from the head region form antennae and mouthparts for sensing the surroundings and processing food. Thoracic appendages are used for walking and for handling food. One or more pairs of abdominal appendages are used in reproduction. *See pages 686–687.*

INTERNAL ANATOMY Lobsters exchange gases using well-protected gills arranged along the cephalothorax. Gases and nutrients are transported around the body in blue blood, which is pumped through an open system that bathes the internal organs. The lobster's massive tail muscles carry out escape responses. *See pages 688–689.*

NERVOUS SYSTEM Giant nerve cells in the ventral nerve cord transmit nerve impulses at high speed to trigger the lobster's escape responses in times of danger. *See page 690.*

REPRODUCTIVE SYSTEM Lobsters have a complex life cycle with a prelarval stage, three planktonic larval stages, and a postlarval stage, followed by a juvenile stage that gradually transforms to an adult through a series of molts. Mating occurs when the adult female molts. *See page 691.*

External anatomy

CONNECTIONS

COMPARE the body structure of a lobster with that of a ***CRAB***. Both kinds of organisms are decapods, and both can regrow lost body parts, such as limbs.

The cuticle of crustaceans is like a hard shell. It serves both as a tough protective layer and as an exoskeleton (outer skeleton) that supports the body. It has joints, where the cuticle is much thinner, to allow movement.

Most American lobsters are predominantly olive green or greenish brown in color, owing to pigment deposited in the cuticle. Their color varies enormously, however, depending on the distribution of patches of orange, red, dark green, black, or blue. This patchwork of color helps to provide camouflage by breaking up the animal's outline against the rocky sea bottom where it lives.

A lobster's body, like that of other crustaceans, is divided into segments. The body segments are most visible on the abdomen, which is the rear body region. The head is fused with the thorax to form a cephalothorax. A structure called the carapace, which is an extension of the exoskeleton, encloses the head and forms a single protective tube of armor plating around the cephalothorax.

A lobster's head bears two pairs of antennae. The small first pair, called antennules, are each branched in two. These antennae are highly sensitive to chemicals and are used to "smell" the surroundings. The second, much larger pair of antennae are almost as long as the lobster's entire body. These are highly sensitive to touch, movement, and vibration. Lobsters hunt at night and can maneuver their antennae to navigate a path in the darkness and detect potential prey or predators in the vicinity.

The lobster's eyes are on movable stalks and provide a very wide field of vision. Each eye contains thousands of light-sensing units, each with a lens. This gives the lobster an image

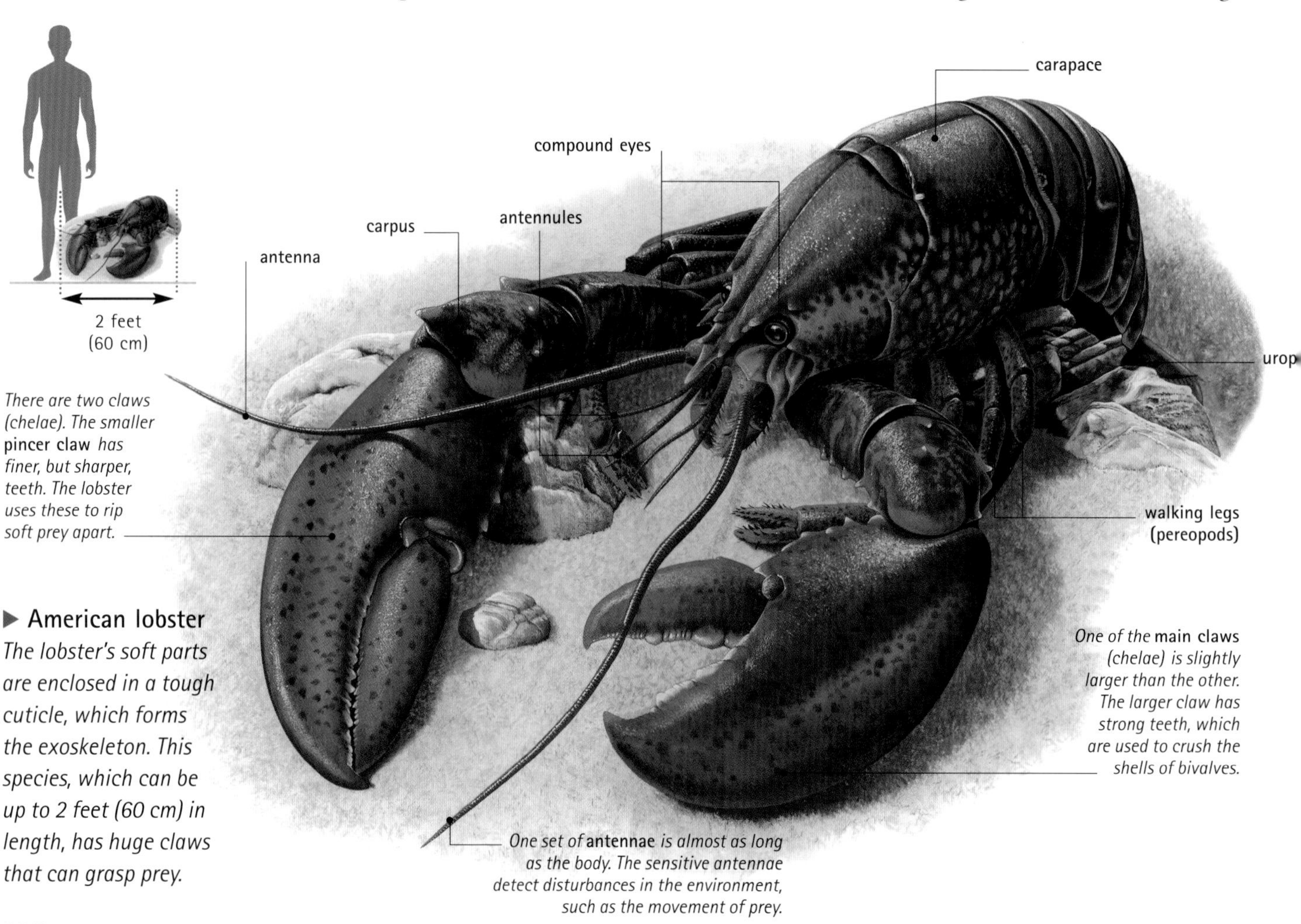

▶ **American lobster**
The lobster's soft parts are enclosed in a tough cuticle, which forms the exoskeleton. This species, which can be up to 2 feet (60 cm) in length, has huge claws that can grasp prey.

made up of thousands of patches, rather like large pixels on a computer screen. The eyes operate in dim light and are most effective at detecting movement, rather than detail.

In more primitive crustaceans, all the appendages of the thorax and abdomen are more or less alike. That arrangement is seen in modern-day brine shrimp. The appendages are used for swimming. In lobsters and other decapods, there are fewer appendages than in the primitive crustaceans, and those in different regions are adapted for different roles. The first three pairs of thoracic appendages are used for handling food. The remaining five pairs on the thorax are longer and stronger and suited for walking, defense, and gathering food. The first pair of walking legs, called chelipeds, have massive, pincerlike claws, called chelae.

On the abdomen, all but one pair of appendages are swimmerets. However, they are too small to propel the lobster's heavy body during swimming, and instead are used in reproduction. The male uses his first pair of swimmerets to transfer a packet of sperm to the female during mating. The female later carries her eggs and young attached to her swimmerets. The last pair of abdominal appendages, the uropods, are broad and branched in two. They function as rudders or levers for swift movement. When threatened, the lobster rapidly flexes its tail under its body, making the whole animal scoot backward, away from danger. In females, the uropods, together with the last body segment (the telson), also form a protective cover over the eggs or young.

IN FOCUS

Different-size claws

One of the lobster's claws, or chelae, usually the left one, is slightly larger than the other claw. The larger claw's serrations and thick armor enable the lobster to crush the hard shells of clams and other invertebrates. The smaller claw has finer teeth, which are used to tear prey animals into manageable chunks. These chunks are then passed to the small pincers on the second and third pairs of legs, and from there to the mouthparts.

CLOSE-UP

The cuticle

The cuticle of lobsters is made of chitin (a complex carbohydrate similar in structure to cellulose), bound up with various proteins. The cuticle is nonliving but is laid down by a single layer of living cells just beneath it called the epidermis. The epidermis produces an outer, colored layer of the cuticle, with a calcium-rich layer beneath it and a calcium-poor layer beneath that. Below the epidermis, integument glands secrete wax through long ducts onto the outer surface of the cuticle, creating the epicuticle, a waxy, waterproof coating. Lobsters need to shed their cuticle once or twice a year to allow growth. The new cuticle takes several months to grow under the old one that is eventually shed.

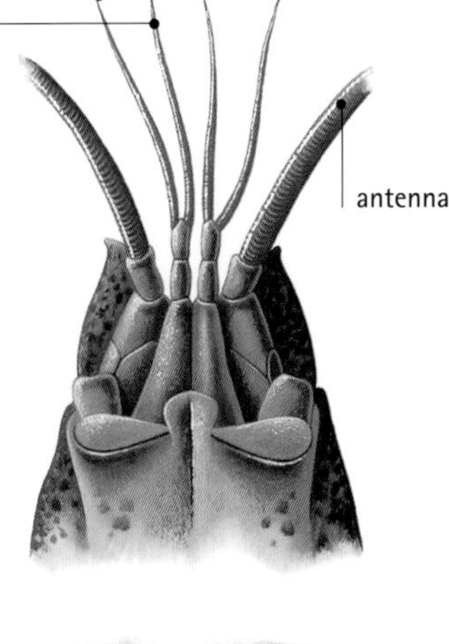

▶ ANTENNAE
American lobster
The inner, branched pair of antennae (called antennules) detect smells. The outer pair are much longer, and are used as feelers.

▶ TAIL
American lobster
The lobster's tail is is made up of sections, which enable it to bend and flex rapidly.

Internal anatomy

CONNECTIONS

COMPARE the diet of lobster larvae with that of a *CRAB*, a *GIANT CLAM*, and a *STARFISH*. All feed on tiny floating organisms called plankton.

Most of a lobster's internal organs lie in the cephalothorax, the front part of the body containing the head and thorax. The circulatory system, as in insects and other arthropods, is an open system. The single-chamber heart pumps blood through arteries into blood spaces that bathe the muscles and other internal organs. The blood drains back to the heart through several holes in the wall of the heart, called ostia. The ostia contain valves that open when the heart relaxes and close when the heart contracts, thereby ensuring that blood flows only one way.

Lobsters, like other decapods, are blue-blooded. This trait is due to the presence of the oxygen-carrying substance hemocyanin, which contains copper and has a bluish hue, as distinct from hemoglobin, the red, iron-containing substance found in vertebrate blood.

The lobster gains oxygen and excretes waste carbon dioxide through its gills. These lie within two branchial (gill) chambers in the thorax. Water enters the chambers through pores between the legs, passes over the gills, and then passes out through two pores in the lobster's head. Hairlike structures (setae) at the bases of the legs filter the incoming water to keep out large particles that could block the breathing apparatus. The water current is maintained by rhythmic beating of a structure called the gill bailer. The lobster reverses the current every few minutes to flush out debris that collects in the gill chambers.

The lobster has 20 gills on each side of the thorax. The gills are arranged in a row alongside the bases of the last two maxillipeds (appendages behind the mouth that help with feeding) and the five walking legs, and extend up into the branchial cavity. Each gill bears numerous feathery filaments, which increase the surface area available for gas exchange. Channels running through the central base that supports each gill carry blood to and from the gills' respiratory surface.

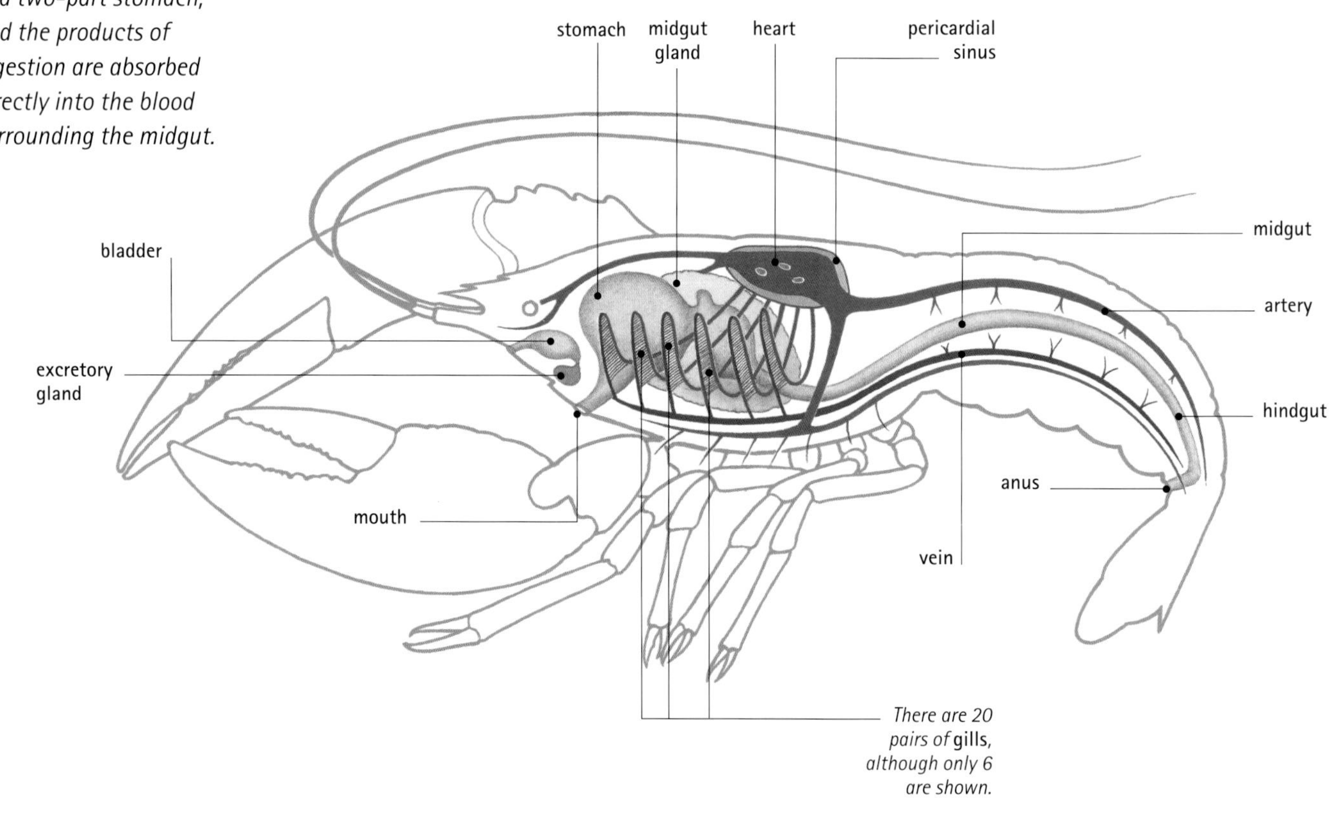

▼ *Compared with that of vertebrates, the lobster heart is very simple. It pumps blue, rather than red, blood. Digestion is carried out in a two-part stomach, and the products of digestion are absorbed directly into the blood surrounding the midgut.*

▲ *An American lobster—a species that lives along the eastern seaboard of North America—comes to grips with a meal of starfish.*

Feeding and digestion

The lobster manipulates food items with its maxillae (accessory mouthparts) and maxillipeds, then crushes them with its mandibles (main jaws) before swallowing the meal into the esophagus. From there, food passes into the two-part stomach. The first part, the cardiac stomach, contains barbed bristles that, together with contractions of the stomach wall, help break up the food. At the entrance to the second chamber, the pyloric stomach, is the gastric mill; this contains toothlike denticles that further grind up the food. The macerated food then enters the pyloric stomach. Small particles pass through a hairlike filter into the midgut glands, where they are chemically digested. Larger particles remain in the pyloric stomach and then pass into the straight section of midgut. The products of digestion are absorbed from the midgut glands and midgut into the surrounding blood. Any waste that remains is packaged into mucus-coated fecal pellets before entering the hindgut. The pellets are finally expelled through the anus. In addition, a pair of glands in the head, called the antennal or green glands, excrete waste substances such as ammonia and excess salts. Soluble wastes also leave the body via the gills.

The lobster's abdomen contains little more than muscle, intestine, the ventral nerve cord, and an artery. The muscle-packed abdomen, or "tail," is the white lobster meat that people favor. Above the intestines lie muscles that extend the abdomen when they contract. Below lie the giant flexor muscles that bend the lobster's tail and produce its escape responses.

Nervous system

CONNECTIONS

COMPARE the structure and function of a lobster's eye with that of a ***HUMAN*** and an ***OCTOPUS***. All use lenses to focus light, but the arrangements of light-detecting cells differ.

A lobster's cerebral ganglion is a collection of nerve cells inside the head that is equivalent to the crustacean's brain. The cerebral ganglion is effectively in two parts: one part is above the esophagus, and the other is below. The two parts are connected by numerous nerve cells. The upper part of the ganglion receives information relating to the major senses through nerves and coordinates overall body movement and behavior. The lower part controls the mouthparts.

Lobsters, like other decapod crustaceans, have a double ventral (front) nerve cord, in contrast to the single dorsal (back) nerve cord of vertebrates. The crustacean nerve cord contains a ganglion (nerve cluster) in each body segment, which connects to nearby internal organs, appendages, and muscles. Relative to the vertebrate nervous system, processing power in the lobster is more decentralized, with many local centers instead of a single brain inside the head.

◀ *The "brain" of a lobster, the cerebral ganglion, is made up of two clusters of cells joined by a collar of nerve cells. The cerebral ganglion connects to a nerve cord running the length of the body.*

The compound eyes of an adult lobster, like those of crabs, have movable stalks. Each eye contains up to 13,500 units called ommatidia, each with a lens, so the lobster's vision is formed from thousands of separate miniature images. The lobster's eyes are adapted to seeing in dim light, and they are temporarily blinded in bright light. They see in black and white, and are particularly good at detecting movement. Shaped like a ball, with the light-sensitive cells on the outside, each compound eye gives the lobster a very wide field of vision so that it can see things approaching from above and below, as well as from in front.

The first antennae, or antennules, are the lobster's organs of smell. When the antennules flick up and down, they are "sniffing" the water. The large antennae behind the antennules are mainly organs of touch. They sense vibrations and currents in the water, which can indicate the presence of nearby prey or predators.

Lobsters have taste receptors on their legs as well as in the mouthparts. This arrangement allows food that is tainted in some way to be quickly rejected. Hairs on the body cuticle are sensitive to water movement and vibrations in a manner similar to the lateral line of fish. At the base of a lobster's antennules lie the statocysts. These function in a way similar to the semicircular canals of many vertebrates, detecting body movement and giving the lobster information about the orientation of its body relative to gravity.

PREDATOR AND PREY

To jump or flip?

When a lobster is threatened by attack from the front, it jumps backward. Giant nerve cells at the top of the ventral nerve cord rapidly relay messages back from the brain to the muscles of the abdomen, triggering the escape response. When the lobster is threatened from behind—which is its weak spot—a second set of giant nerve cells in the middle of the nerve cord sends nerve impulses only to the first three abdominal segments. This produces a different response. The lobster flips forward and somersaults, first turning on its back and then righting itself to confront the danger.

Reproductive system

Lobsters have separate sexes—male and female. The female's two ovaries (egg-producing organs) lie inside the cephalothorax but extend deep into the abdomen. The female releases her ripe eggs along a pair of oviducts that run from the ovaries to an opening at the base of the third walking legs. The male's two testes (sperm-producing organs) lie inside the cephalothorax and release sperm through two sperm ducts that extend to the base of the fifth walking legs. Sperm passing along the ducts are coated in a gel that glues them together into a spermatophore, or sperm packet. During mating, the male transfers a spermatophore to the female using his first pair of swimmerets.

The behavior of lobsters and crabs is among the most complex of all invertebrates. They use body posture, and movements of their chelae, walking legs, and antennae to signal to rivals or mates, either to settle territorial disputes or in courtship before mating. American lobsters mate after a courtship ritual, which stimulates the female to shed her old cuticle. While the female is recovering from the molt, the male turns her over, grasps her with two copulatory appendages, and deposits a sperm packet inside a cavity, called the seminal receptacle, on her underside. The eggs are fertilized when they pass over the sperm packet.

▲ *The eggs produced by a female lobster remain attached to the mother's body for several months.*

A few days later, the female releases thousands of fertilized eggs that attach by a gluelike substance to her swimmerets. The eggs develop there for 9 to 12 months until they reach a prelarval stage, when they hatch and are wafted into the water by movements of the mother's tail. As they rise to the surface, the prelarvae molt into first-stage larvae. These molt through two more larval stages over two to seven weeks, consuming larger plankton (floating organisms) as they grow. At the fourth molt, a postlarva emerges and swims to the seabed to find a suitable bottom layer on which to settle. The postlarva molts into a juvenile, which is like a miniature version of the adult. Lobsters take 5 to 8 years to grow to adult size, and may live more than 50 years.

TREVOR DAY

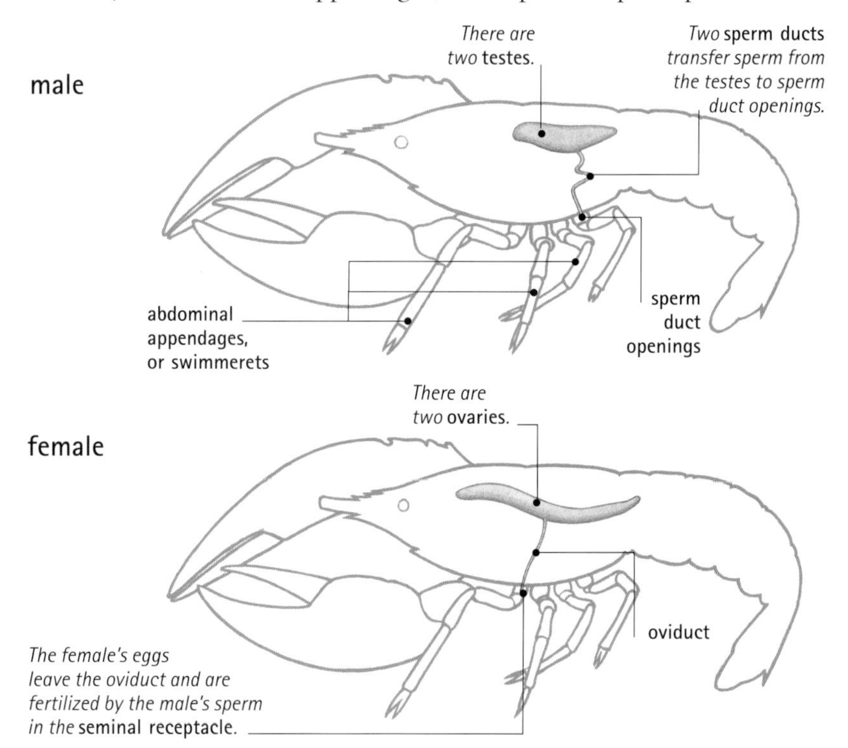

▼ **REPRODUCTIVE ORGANS**
A male lobster has two genital openings and two copulatory appendages on its ventral surface (underside). During mating, the male uses the appendages to transfer a sperm packet from the sperm duct opening to the female's seminal receptacle. There, the eggs are fertilized. The eggs then attach to the female's spinnerets until they hatch.

FURTHER READING AND RESEARCH

Factor, J. R. 1995. *The Biology of the Lobster, Homarus americanus*. Academic: New York.

Ruppert, E. E., R. S. Fox, and R. B. Barnes. 2004. *Invertebrate Zoology: A Functional Evolutionary Approach*. 7th edition. Brooks Cole Thomson: Belmont, CA.

The Lobster Conservancy: www.lobsters.org

Louse

CLASS: Insecta ORDER: Phthiraptera

Lice are small, wingless insects. They feed on the blood and skin of mammals and birds and feathers. Lice have a flattened body equipped with strong claws to help them cling to their hosts (the animals on which they feed).

Anatomy and taxonomy

Scientists group organisms into taxonomic groups based largely on features of their anatomy, called morphological characters. Genetic characteristics are also used to classify organisms by showing how closely related two species are on the basis of the similarity of their DNA. Sometimes, organisms that have similar names may not actually be closely related, because scientists originally named them on the basis of their similar appearance or behavior.

• **Animals** All animals are multicellular and feed off other organisms. They differ from other multicellular life-forms in their ability to move around (generally using muscles) and to respond rapidly to stimuli.

• **Arthropods** These are invertebrate animals with jointed legs. Instead of having an internal skeleton, the body is supported by a tough outer covering called an exoskeleton. There are three main divisions of arthropods: Chelicerata (scorpions, spiders, mites, and ticks), Crustacea (mainly crabs and shrimps), and Hexapoda (insects, springtails, and other six-legged arthropods). Wood lice—also called pill bugs—are also crustaceans, but they are not true lice. Most relatives of wood lice are aquatic, but wood lice live on land. Unlike the true lice, wood lice are not parasites. Instead, they live beneath stones and wood, and they feed on decaying plant matter.

• **Hexapods** These are six-legged arthropods. The vast majority of hexapods are insects, but there are three small noninsect hexapod groups. Noninsect hexapods do not have wings, and they differ from the insects in the structure of their mouthparts. Noninsect hexapod mouthparts are kept in a pouch on the underside of the head and are everted (popped out) for feeding.

• **Insects** The insects display the most diverse body forms and functions of all the classes of animals on Earth. However, all insects have three pairs of segmented legs, and their body is organized into a head, thorax, and abdomen. The class Insecta contains at least 26 orders, among them the moths and butterflies (Lepidoptera), flies (Diptera), beetles (Coleoptera), dragonflies (Odonata), aphids (Homoptera), and true lice (Phthiraptera).

• **Lice** The order Phthiraptera contains around 3,500 species of true lice. Lice live permanently as ectoparasites (external parasites) of birds and mammals. Traditionally, taxonomists have placed the true lice in two suborders, Anoplura (sucking lice) and Mallophaga (chewing lice), based largely on lifestyles, but more recent research recommends a division into four groups.

• **Amblycera** This suborder contains the most primitive species of lice. Their mouthparts have both mandibles (jaws) and maxillary palps (appendages), and they feed

▶ *The true lice belong to the order Phthiraptera, one of at least 26 orders of insects. Some of the relationships of the true lice are shown in the family tree.*

▲ *This electron microscope image is of a human head louse (a species of sucking louse) and one of its eggs. The image magnifies the insect around 100 times.*

primarily by chewing feathers or skin. Some chewing lice—for example, the shaft louse (*Menopon gallinae*)—infest poultry and cause serious damage to affected birds.

• **Ischnocera** Most of these lice are parasites of birds, although those in the family Trichodectidae infest mammals. They feed on feathers, hair, or skin. Their mouthparts have mandibles but no maxillary palps.

FEATURED SYSTEMS

EXTERNAL ANATOMY Lice are wingless insects with a flat body. They have six legs, each with one or two claws. *See pages 694–695.*

INTERNAL ANATOMY Lice have small compound eyes and short antennae, which bear sensory organs. They breathe through openings called spiracles on their back. *See page 696.*

DIGESTIVE AND EXCRETORY SYSTEMS Lice have mouthparts that differ in structure according to their diet. Some have chewing mouthparts to eat feathers, hair, or skin, while others have sucking mouthparts to drink blood. *See pages 697–698.*

REPRODUCTIVE SYSTEM Male and female lice mate on their hosts. Females attach their eggs to the host's body. *See page 699.*

• **Rhyncophthirina** This suborder contains only two species of lice, both of which have mandibles at the end of a long snout. The snout helps them cut into the thick skin of their hosts—large mammals, including elephants and warthogs—to obtain a blood meal.

• **Anoplura** The 500 or so species in the suborder Anoplura have highly specialized mouthparts for sucking blood. In addition to the family Pediculidae, the Anoplura suborder contains the family Phthiridae, the best-known example of which is the pubic or crab louse, *Phthirus pubis*, which inhabits the pubic region of humans.

• **Pediculidae** The human head louse, *Pediculus humanus capitis*, is one of two subspecies of human louse, the other being *Pediculus humanus corporis*, the human body louse, which is slightly larger. Unlike the body louse, which is a vector (carrier) of diseases such as typhus, the human head louse is not known to transmit any diseases. However, it does cause itching and skin inflammation called pediculosis.

External anatomy

CONNECTIONS

COMPARE the claws of a louse with those of a ***CRAB***. Crabs use their claws for defense, display, and feeding.

COMPARE the exoskeleton of a louse to the skeleton of a vertebrate animal, such as a ***TROUT***. The louse's body is supported from the outside rather than by internal bones.

The human head louse is 0.04 to 0.08 inch (1 to 2 mm) long. Like other lice, it has a body flattened from top to bottom; this shape, called dorsoventral flattening, helps the louse hide among its host's hairs. Fleas, which are also parasites of birds and mammals, are flattened from side to side (laterally).

Most lice have a small head with one pair of eyes and a pair of short antennae. Behind the head is the thorax, which is formed from three segments: the prothorax, mesothorax, and metathorax. In anopluran lice, these three sections are fused to form one large thoracic segment. In the other three suborders, the prothorax is separate from the other two segments. Behind the thorax, the abdomen forms the the rear of the body.

Lice have three pairs of short, strong legs, one for each thoracic segment. The legs end in one or two strong claws, depending on the species. The head louse has only one claw on each leg, which meets a thumblike structure on the side of the last leg segment. The claws

IN FOCUS

Hitching a ride

Lice have no need for wings because they spend most of their life on the same host, occasionally transferring if hosts come into close contact, as when schoolchildren brush heads in the playground. If a bird carrying ischnoceran lice dies, the lice take the risk of hitching a ride on louse flies. The lice attach themselves by their mandibles to the fly's abdomen and drop off when it reaches another bird. Traveling on another organism, without causing it harm, is known as phoresy. Other parasitic organisms also have inventive methods of traveling between hosts. Fleas, for example, have powerful legs to jump from host to host; and ticks stand on the tops of grasses, waving their legs, until a host walks by.

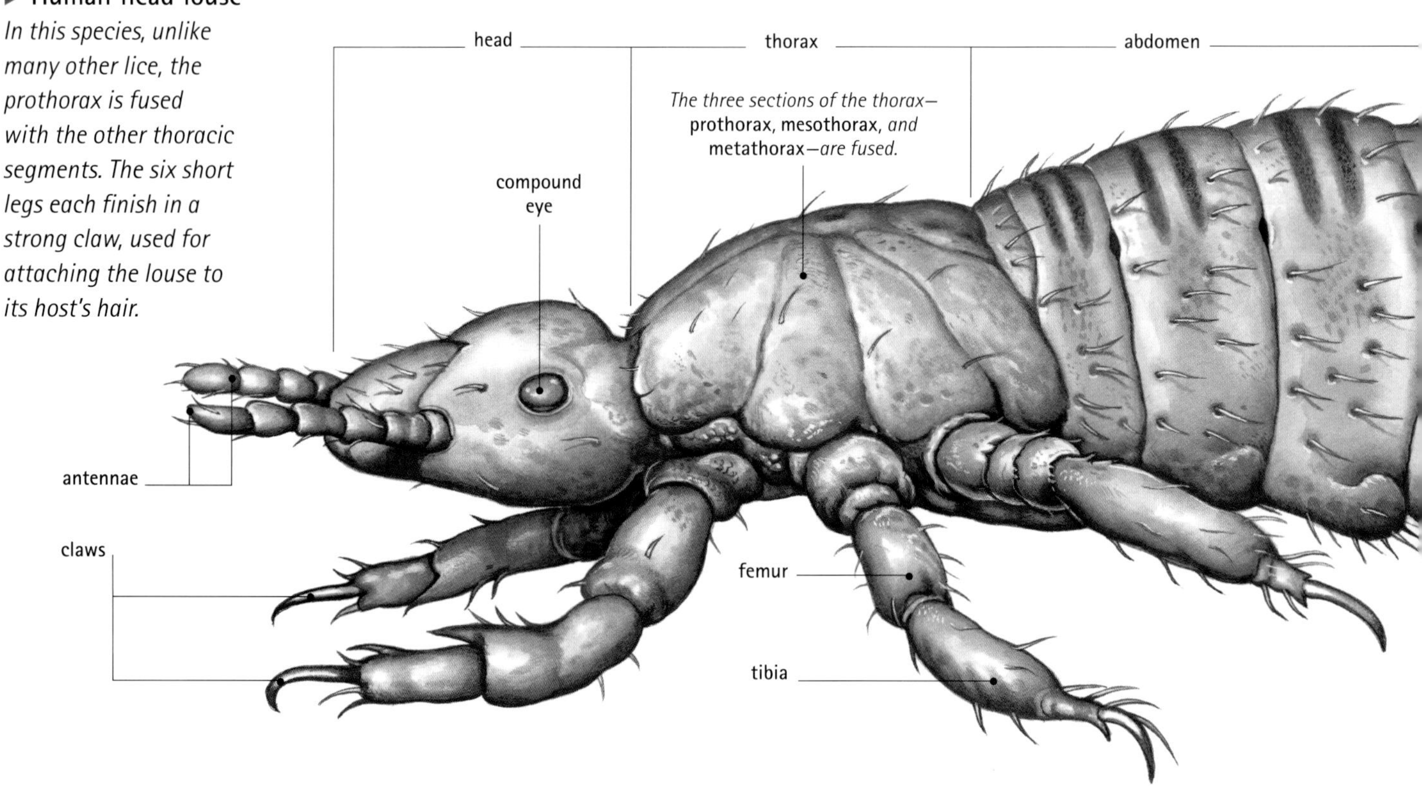

▶ **Human head louse**
In this species, unlike many other lice, the prothorax is fused with the other thoracic segments. The six short legs each finish in a strong claw, used for attaching the louse to its host's hair.

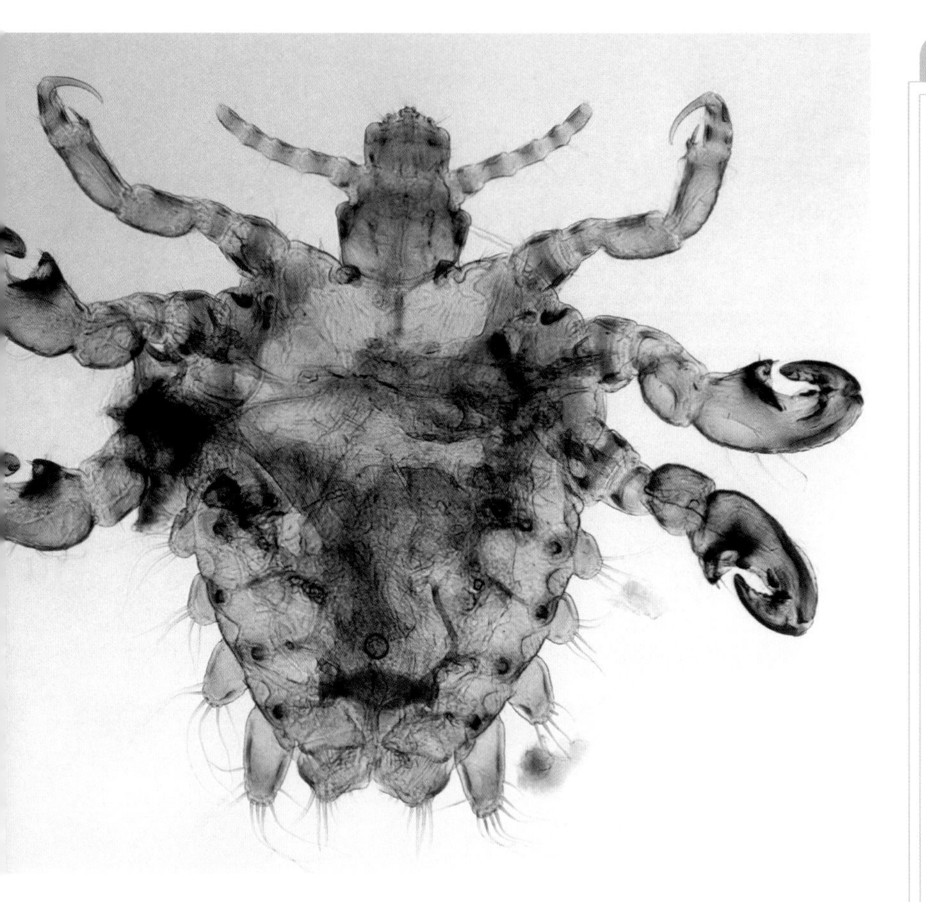

▲ *The crab louse,* Phthirus pubis, *lives in the pubic hair of humans. The hairs in the pubic area have a flat cross section. The louse's claws have evolved a form that enables the louse to firmly grip these hairs.*

GENETICS

Lice and DNA studies

Analysis of the similarities and differences in DNA of different species can be used to estimate when they diverged from a common ancestor. By examining the DNA of the human head louse and body louse, German scientists have estimated that these two subspecies diverged around 72,000 years ago. The scientists suggested that the trigger for the split might have been the first use of clothes by humans at around that time. Before this, there was little difference between living in the hairs on the head and living in hairs on the body, but suddenly lice had a new niche to exploit—the humid, sheltered environment inside clothing.

Analysis of louse DNA has also been useful in discovering the evolutionary relationships between other animal species. By comparing lice from chimpanzees with lice from humans, scientists have shown that the different louse species diverged around 5.5 million years ago—about the same time as the host species themselves. Similarly, the relationship between flamingos, storks, and ducks was made clear by looking at their lice. Although flamingos superficially resemble storks, they are actually more closely related to ducks, with which they share similar species of lice.

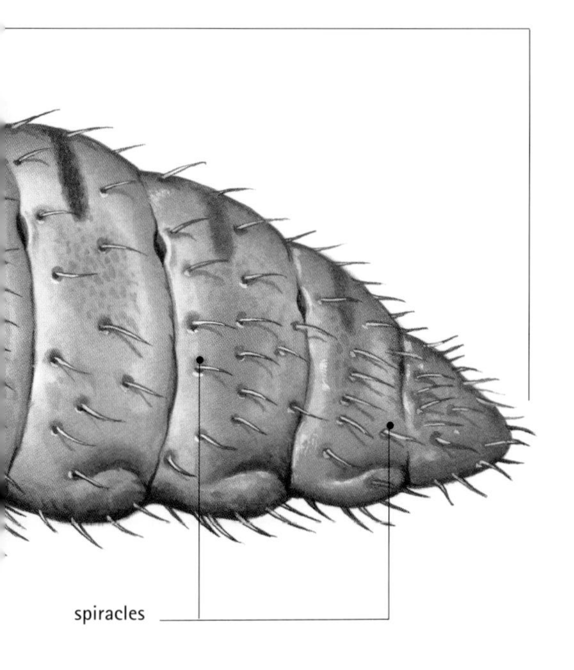

allow the louse to grip six hairs simultaneously, making it very difficult to dislodge. The claws differ in shape between species, with each being adapted to the shape of the hair onto which it must grasp. Head lice and body lice, for example, can grip hairs with a circular cross section, such as those on the head and body, but cannot grip hairs with a flat cross section, such as those in the pubic region.

Tough exoskeleton

Like all insects, lice are covered with a tough outer coating called an exoskeleton, or cuticle. It covers the entire body surface and also lines the digestive tract and the airways, which open on the outside at the spiracles. The exoskeleton provides support for the body, which lacks an internal skeleton, and minimizes water loss.

Internal anatomy

CONNECTIONS

COMPARE the nervous system of the louse with that of the ***HUMAN***. Both are made up of nerve cells (neurons), but in humans the central nervous system is far more complex.

Lice, like other insects, have muscles, nerves, a circulatory system, a respiratory system, and a digestive sysem. Like most animals, lice use muscles within their legs for locomotion. They can move quickly among hairs and feathers, but are surprisingly slow and clumsy when traveling on other surfaces. Louse "feet" are made for grasping, not for walking.

The circulatory system of a louse comprises a body cavity called a hemocoel, which is filled with a bloodlike fluid called hemolymph. The hemolymph bathes the cells of the internal organs, providing them with nutrients and oxygen and removing the waste products of metabolism.

Like most insects, lice breathe through spiracles, which are porelike openings on the surface of the thorax and abdomen. The spiracles can be closed to reduce water loss from the louse's body and also to prevent liquid from entering if the louse is immersed in water. Inside the louse's body, the spiracles open into tubes called tracheae. These tubes provide the tissues beneath the surface with a large surface area for gas exchange. Lice that live on seals have overlapping, flattened bristles called setae on their body; these trap a bubble of air so that gas exchange can continue even when the seal is submerged underwater.

CLOSE-UP

Sensing their environment

The antennae of the head louse have several types of sensory organs that perceive changes in the environment. Retractable peg organs detect smells, tuft organs monitor humidity, and fine hairs sense touch. It is important that the louse senses changes in its environment—for example, an increase in temperature or humidity might indicate that the host animal has a fever and could die. The louse should therefore take the first opportunity to move to a new host.

Nervous system

Insects have pairs of ganglia in each body segment. Each ganglion contains a cluster of nerve cells connected by axons (nerve fibers) that link together into a nerve cord that runs the length of the body. When a louse receives a stimulus—for example, a touch to its body—electrical impulses travel from the sense organ (in this case a hair cell) to the nerve cord along sensory neurons (nerve cells). The ganglia process the incoming impulse and send out an impulse along other neurons connected to muscles to stimulate an appropriate response. This might, for example, cause the louse to grasp a hair tightly with its claws to prevent being brushed off.

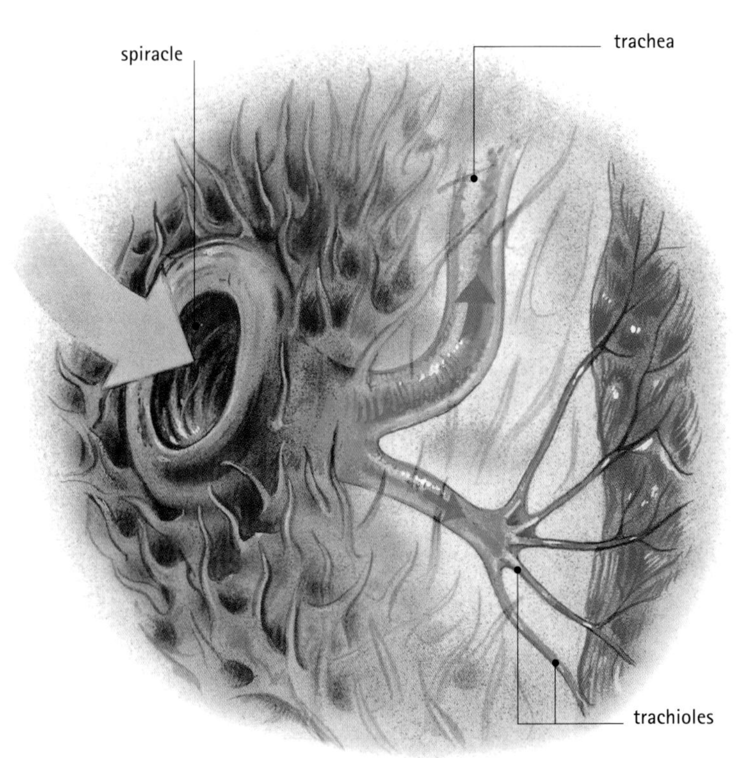

◄ TRACHEAE

The body of a louse is covered with many porelike spiracles. Air enters through the spiracles and passes into the tracheae, a series of tubes stiffened with tough rings. The tracheae branch into smaller trachioles, which end within the tissues. There, gaseous exchange takes place.

Digestive and excretory systems

The mouthparts and digestive tract of the human head louse are suited to a diet of blood. Lice are able to feed immediately when they hatch, and from this moment onward they must obtain a meal several times a day.

The basic structure of insect mouthparts includes a labrum (equivalent to an upper lip), a labium (equivalent to a lower lip), paired mandibles and maxillae (jaws) for chewing and manipulating food, and the hypopharynx, which contains the salivary duct. These parts take different shapes in different insects, depending on the the way they feed, and not all parts are present in louse species. For example, the mouthparts of the head louse are used to pierce the skin of its host and suck up a blood meal. Consequently, the head louse does not have any mandibles, because it does not need to chew its food. The remaining mouthparts form a long proboscis-like structure, which is retracted into the head when the louse is not feeding.

The labrum of a head louse is called the haustellum; it has toothlike structures on its tip, which grip the surface of the host's skin. The haustellum supports a bundle of piercing organs, or stylets, called the fascicle. One stylet is derived from the two maxillae in other insects, and forms a food channel. The second stylet is the hypopharynx through which saliva enters the host's wound; and the third stylet, derived from the labium, has a serrated tip for cutting into the hosts' skin. Lice that feed by chewing feathers or skin generally have simpler mouthparts than those that feed on blood.

Lice have small salivary glands in the head, and two large pairs of salivary glands in the thorax. Ducts transfer the secretions from these glands to the stylets. The saliva causes an allergic reaction in many hosts: hence the unpleasant itchiness associated with louse infestations.

Digestive tract

Blood flows into the louse's gut through a combination of the blood pressure in the host animal's capillaries (smallest blood vessels) and suction-producing structures in the foregut called the pharanx and cibarial pump. The blood enters the midgut, the capacity of which is increased by two side pouches, the gastric ceca. These expand as the louse feeds, and the gut fills with blood. On the wall of the louse's midgut is a disklike structure, called a mycetome. This contains microorganisms, probably bacteria, which live in symbiosis with the lice. Symbiotic organisms receive mutual benefit from one another, and neither organism harms the other. The louse provides

CONNECTIONS

COMPARE the mouthparts of the louse with those of the ***HOUSEFLY***.

COMPARE the excretory system of the louse with the kidneys of the ***HUMAN***.

COMPARE the symbiotic gut bacteria of the louse with those of the ***WILDEBEEST***.

▼ **DIGESTIVE SYSTEM**

Waste products from the louse's metabolism move along the Malpighian tubules to the hindgut.

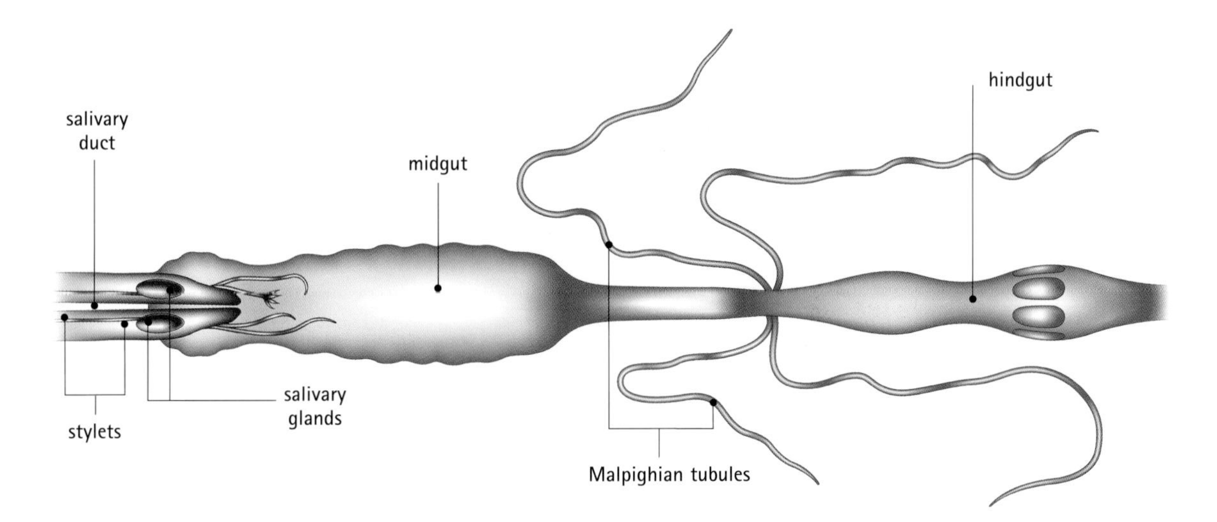

▲ *Human body lice feed on blood taken from their host. The blood is sucked into the gut and digested with the help of bacteria. The blood-filled intestine of this louse is clearly visible.*

the microorganisms with shelter and passes them on to its offspring; the microorganisms probably help the louse to digest the blood and may produce vitamins that the louse does not obtain directly from its diet.

COMPARATIVE ANATOMY

Feeding: lice and ticks

Lice feed intermittently for short periods of time, but this is not the case with some other familiar parasitic animals. In particualr, ticks feed continuously, sometimes for days or even weeks. They attach themselves to their host's skin by a part of the mouth called the hypostome. This part has backward-projecting spines on it, which prevent the tick from being dislodged while feeding. As they feed, ticks increase greatly in size to accommodate the large volume of blood. After feeding, the ticks drop off to find another host or to lay eggs. Unlike lice, some ticks can go for great lengths of time between blood meals.

Excretory system

The removal of metabolic waste, or excretion, occurs through Malpighian tubules located near the junction of the midgut and hindgut. These blind-ended tubes float freely in the hemolymph. Waste products of metabolism—mainly nitrogen-containing compounds such as uric acid—accumulate in the blood and diffuse into the tubules. From there the waste passes into the hindgut, which has six rectal papillae (projections). These function like kidneys, reabsorbing water from the waste. The resulting highly concentrated mixture of uric acid and dry feces is ejected from the digestive tract through the anus.

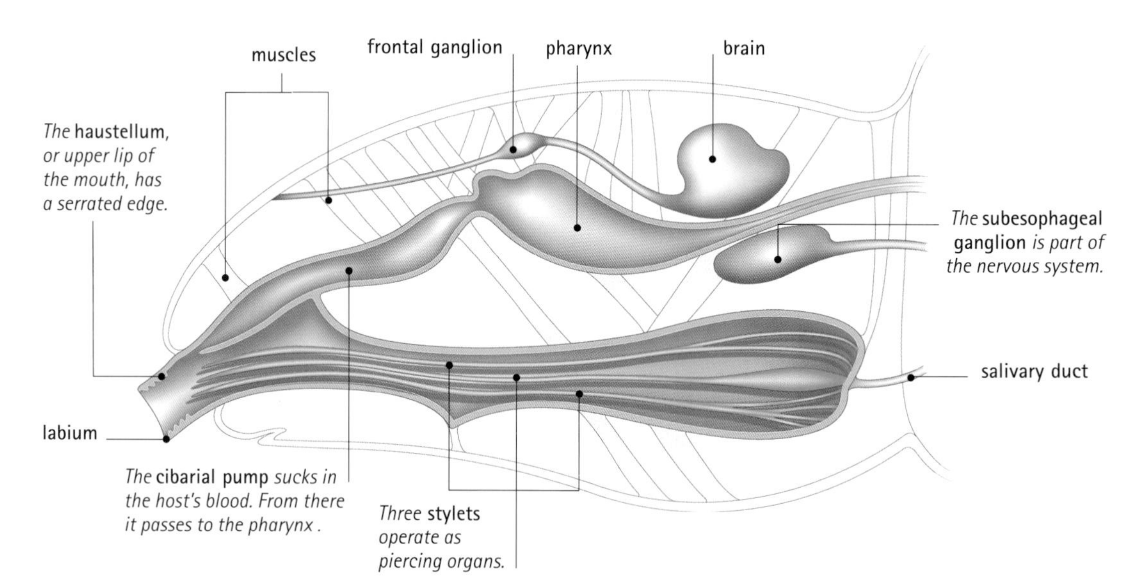

▶ MOUTHPARTS
The stylets pierce the skin of the host, and saliva from the salivary gland is pumped into the wound. The action of the cibarial pump sucks the host's blood into the pharynx.

Reproductive system

Lice reproduce sexually. Male and female lice are distinguishable from one another by the shape of their abdomen. In males, the abdomen is rounded with the anus and sexual opening on the upper surface; this allows the male to mate from beneath the female. In contrast, female lice have a V-shape opening at the end of their forked abdomen. In addition, two families of ischnoceran lice have sexually dimorphic antennae—that is, the antennae of males and females differ in appearance. Those of the male lice in those families are greatly enlarged compared with those of the females.

While mating, the male louse restrains the female, usually by grasping one of her legs with his claws. The male produces a penislike structure called an aedeagus from his sexual opening, or cloaca. The aedeagus is connected to a vesicle; this expands, forcing the aedeagus out. The aedeagus is then inserted into the vagina of the female louse, and a packet of sperm called a spermatophore is transferred into the vagina through the aedeagus. The sperm fertilizes the female's eggs.

Cementing eggs

Female lice lay their eggs a few days after mating. Head lice cement their eggs, called nits, to the base of hairs, to prevent them from being dislodged by the host animal. The eggs have a hatchlike opening, called an operculum, for gas exchange. Empty nits from which lice have hatched are often the first indication of a louse infestation because they are pale in color and thus more visible than the adult lice. Body lice glue their eggs to clothing, whereas lice that infect birds attach their eggs to the base of feathers. Female lice need to mate again before they can lay another batch of eggs. A single female human head louse may lay around 300 eggs in its lifetime.

KATIE PARSONS

IN FOCUS

Metamorphosis

Lice have a hemimetabolous life cycle: that is, young lice undergo a process of gradual change between hatching and eventual maturity. They pass through three larval stages, known as instars, becoming more like the adult at each stage. Development from egg to adult may take between three and four weeks. However, human head louse larvae become sexually mature only eight days after hatching.

FURTHER READING AND RESEARCH

Ruppert, Edward E., and Robert D. Barnes. 1994. *Invertebrate Zoology*. Saunders College: Fort Worth, TX.

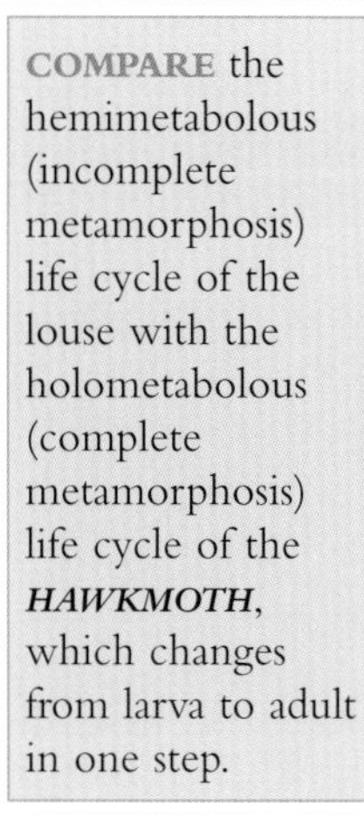

COMPARE the hemimetabolous (incomplete metamorphosis) life cycle of the louse with the holometabolous (complete metamorphosis) life cycle of the ***HAWKMOTH***, which changes from larva to adult in one step.

▲ MALE REPRODUCTIVE ORGANS

▲ FEMALE REPRODUCTIVE ORGANS

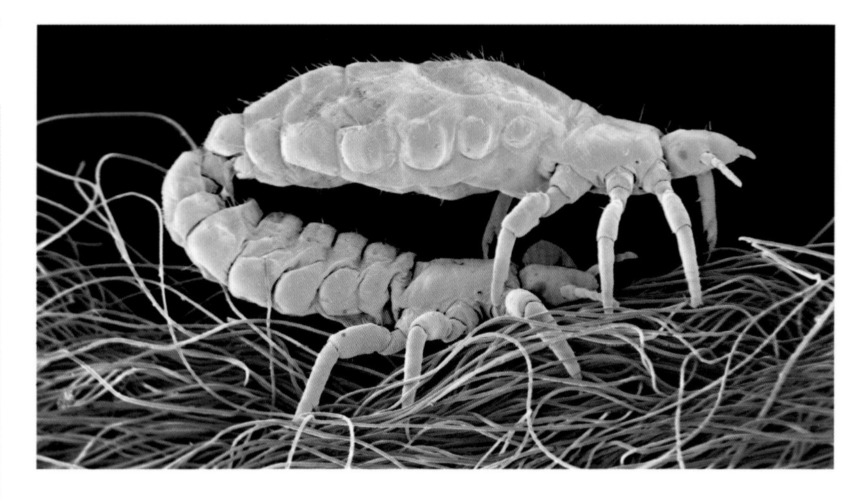

▲ *Human body lice mate using structures at the ends of their abdomen. The male is slighly smaller than the female, and is positioned beneath her.*

Manatee

ORDER: Sirenia FAMILY: Trichechidae
GENUS: *Trichechus*

The three species of manatees live in separate geographical locations. The West Indian manatee inhabits the warm coastal waters and rivers of the southeastern United States, Central America, and the islands of the Caribbean; the Amazonian manatee lives in northern South America; and the West African manatee is found in western Africa. These large, sluggish, aquatic mammals are vegetarian and feed variously on water plants in freshwater, brackish (slightly salty) water, or salt water.

Anatomy and taxonomy

Scientists categorize all organisms into taxonomic groups based partly on anatomy. The three species of manatees, together with the dugong, are sirenians—that is, members of the order Sirenia. A fifth sirenian, Steller's sea cow, became extinct in the 1760s as a result of intensive hunting.

- **Animals** Manatees, like other animals, gain their food by eating other life-forms. Animals differ from other multicellular organisms in their ability to move from one place to another (in most cases, using muscles). They generally react rapidly to touch, light, and other stimuli.

- **Chordates** At some time in its life cycle a chordate has a stiff, dorsal (along the back) supporting rod called the notochord that runs all or most of the length of the body.

- **Vertebrates** In vertebrates, the notochord develops into a backbone (the spine or vertebral column) made up of separate units called vertebrae. The vertebrate muscular system moves the head, trunk, and limbs. It consists primarily of muscles arranged in mirror-image symmetry on either side of the backbone.

- **Mammals** Mammals are warm-blooded vertebrates that have hair made of keratin. Females have mammary glands that produce milk to feed their young. In mammals, the lower jaw is a single bone (the dentary) hinged directly to the skull—a different arrangement from that found in other vertebrates. A mammal's inner ear contains three small bones (ear ossicles), two of which are derived from the jaw mechanism in mammalian ancestors. Mammalian red blood cells, when mature, lack a nucleus; all other vertebrates have red blood cells that contain nuclei.

- **Placental mammals** These mammals nourish their unborn young through a placenta, a temporary organ that forms in the mother's uterus (womb) during pregnancy.

- **Sirenians** Members of this group of mammals are well fitted for life in water, where they spend their entire life. Sirenians, also called sea cows (because of their docile nature and grazing habit), look superficially like a cross

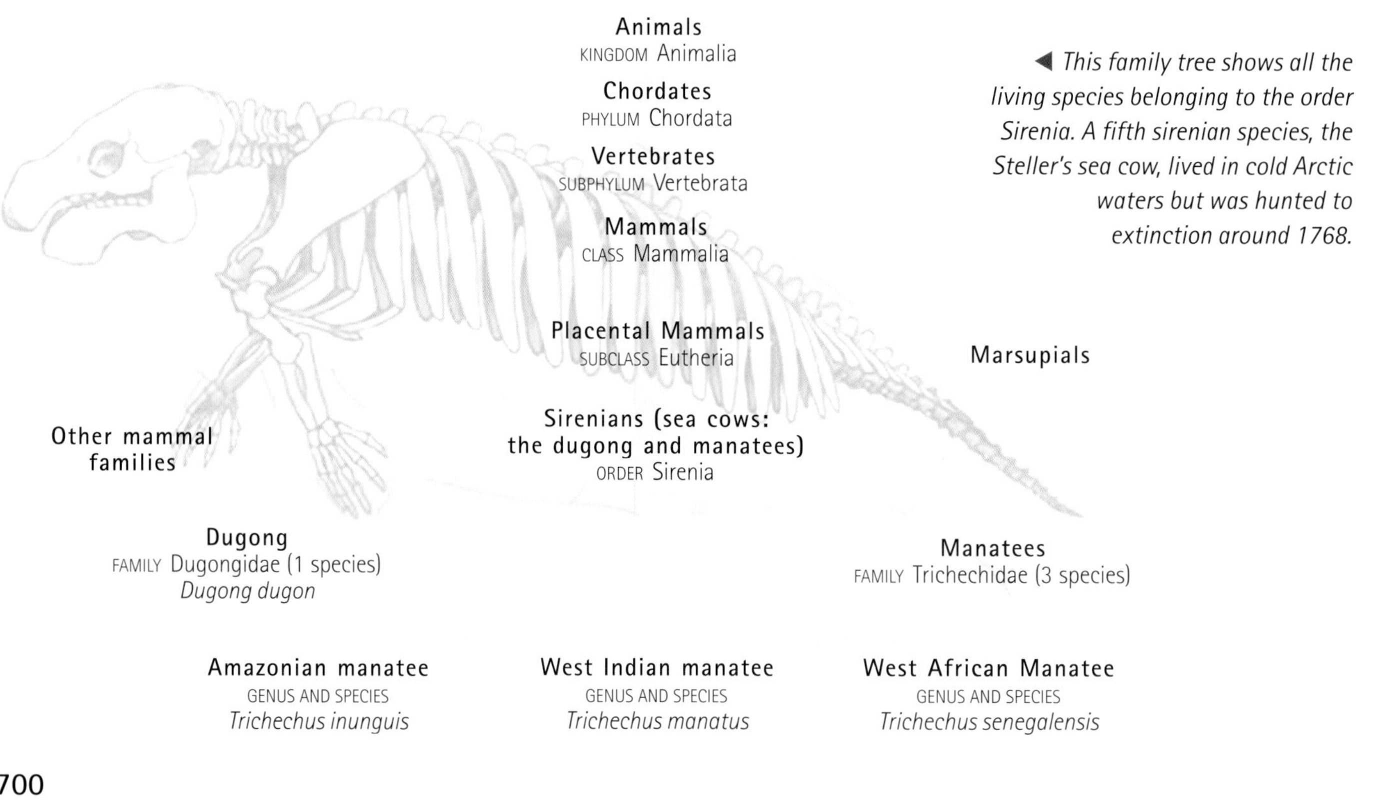

◀ *This family tree shows all the living species belonging to the order Sirenia. A fifth sirenian species, the Steller's sea cow, lived in cold Arctic waters but was hunted to extinction around 1768.*

between a walrus and a small whale, but they are not closely related to either. The name sirenian comes from the myth of sirens, or mermaids, because early sailors mistook these animals for creatures that were half woman, half fish.

Sirenians are slow-swimming and bulky, with a streamlined body shape that reduces drag as the animal swims. Sirenians have a paddlelike tail; flippers that are modified forelimbs with no visible digits (toes or fingers); no hind limbs; and no dorsal fin. Sirenians give birth underwater. They are the only mammals that have evolved to graze plants in coastal waters, and their dentition (teeth) evolved for this purpose.

• **Dugong** Dugongs, unlike manatees, live only in salt water. They are largely restricted to water warmer than 64°F (18°C). The sea grasses on which dugongs feed grow in shallow water, and dugongs rarely dive deeper than 65 feet (20 m). A dugong's tail is broad with a straight or slightly concave trailing edge. The snout hangs down to form a flexible, bristle-covered muzzle. Adult males have incisor teeth that point forward as very short tusks.

• **Manatees** West Indian and West African manatees range between salt water and freshwater, but the Amazonian manatee spends its entire life in freshwater. Some scientists recognize two subspecies of the West Indian manatee: the northerly Florida manatee and the southerly Antillean manatee. The subspecies have distinct geographic distributions and are distinguished largely on the basis of subtle features of the skull and biochemical differences. Manatees have a paddlelike tail, similar in shape to that of beavers. The manatee's muzzle is less downturned than that of dugongs, and manatees have many more teeth. Manatees' teeth form at the back of the mouth and move forward to replace worn teeth at the front.

▲ *The West Indian manatee is the largest of the three living manatee species. It lives in the warm coastal waters of Central America, the southeastern United States, and the Caribbean.*

FEATURED SYSTEMS

EXTERNAL ANATOMY Manatees are sirenians (sea cows) with a relatively streamlined body, paddlelike flippers, and a horizontally flattened tail. *See pages 702–704.*

SKELETAL SYSTEM To provide ballast, the bones are particularly heavy, with the spine acting as an anchor for muscles that raise the tail up and down. *See pages 705–707.*

MUSCULAR SYSTEM A large sheet of muscle extending from head to tail along both sides of the body protects the contents of the chest and abdominal cavities and provides the power stroke in swimming. *See page 708.*

NERVOUS SYSTEM The sirenian brain is unusually small and simple in structure, compared with brains of other marine mammals. In manatees, touch, taste, and hearing are more important than vision and smell. *See pages 709–710.*

CIRCULATORY AND RESPIRATORY SYSTEMS Sirenians do not need high energy levels, so their circulatory and respiratory systems are relatively simple. Manatees have unusually elongated lungs. The lungs, together with a very large diaphragm, enable the animal to make fine adjustments to its buoyancy in the water. *See pages 711–712.*

DIGESTIVE AND EXCRETORY SYSTEMS Sirenians digest plant material using bacteria in their intestines, as elephants and horses do, rather than in their stomachs, as in cows and sheep. *See pages 713–714.*

REPRODUCTIVE SYSTEM The female's mammary glands are located under the flippers, and the male's sex organs are internal. These features aid streamlining. *See pages 715–717.*

External anatomy

CONNECTIONS

COMPARE the mammary glands of a female manatee with the similar arrangement of an ***ELEPHANT***. Both animal's mammary glands are just inside the forelimbs.

COMPARE the flapless ears of a manatee with the ears of an ***OTTER***, which have flaps.

Manatees and dugongs have a basically fusiform (torpedo-shape) body with a head that merges with the trunk and no visible neck. The body shape helps the animals move through the water with minimal drag but is less streamlined than that of a shark or dolpin. Sirenians rarely need to swim fast because they graze on plants rather than hunt fish and tend to rely on their bulk or their numbers to avoid attacks by predators. In addition, many large predators, such as sharks and toothed whales, do not hunt in the shallow waters where sirenians usually graze.

Adult West Indian and West African manatees can grow up to 15 feet (4.6 m) long and weigh 2,500 pounds (1,136 kg). The dugong is slightly smaller, up to about 13 feet (4 m) and 2,000 pounds (900 kg), and the Amazonian manatee smaller still, at 10 feet (3 m) and 1,100 pounds (500 kg). Steller's sea cow was a massive sirenian that grew to 25 feet (7.5 m) long and weighed up to 6.5 tons (5.9 metric tons). It became extinct around 1768.

Accomplished swimmer

Manatees and dugongs swim using up-and-down movements of their flattened tail to drive them forward, in a manner similar to whales. Sirenians steer by flexing the body and tail and adjusting the angle of their forelimbs, which are flattened into paddlelike flippers. The flippers move at the elbow, with the upper

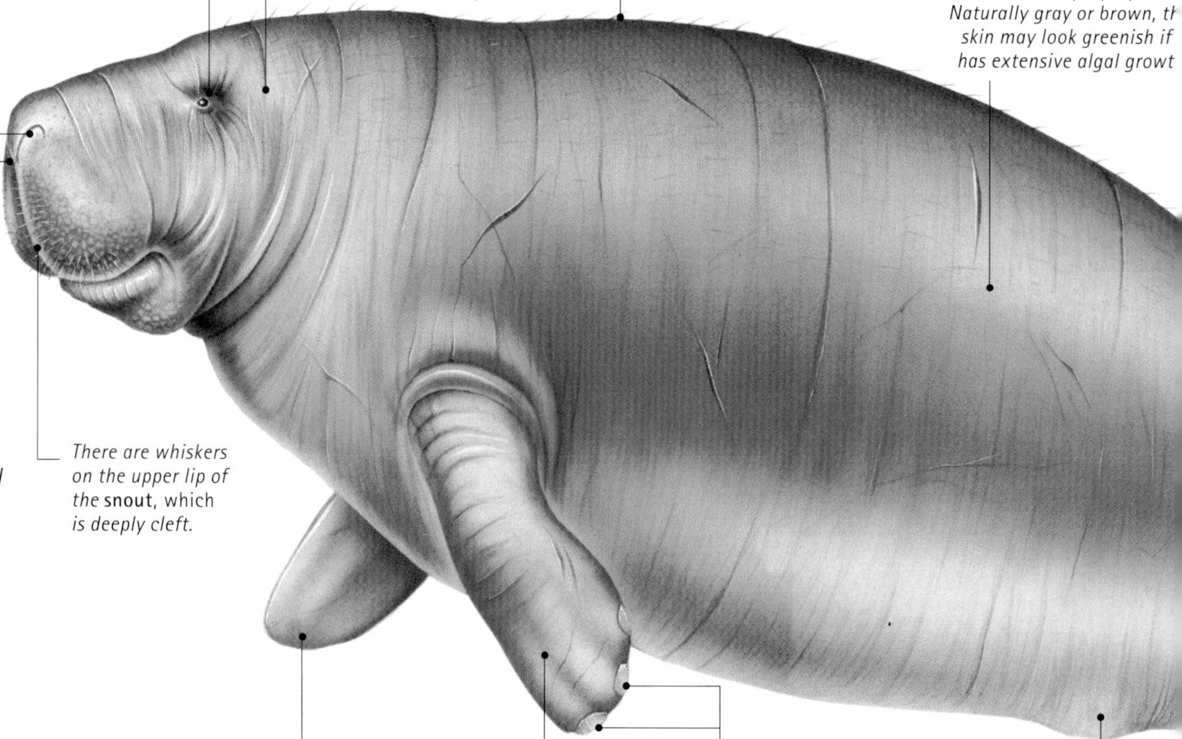

The **eye** *is small in relation to the size of the head.*

The **ear canal** *opens just behind the eye. There is no earflap.*

The skin has a sparse covering of pale **hairs.**

The manatee's **skin** *is roug and is often scarred fro contact with ships' propeller Naturally gray or brown, th skin may look greenish if has extensive algal growt*

The two **nostrils** *can be closed by valves when the manatee dives.*

▶ **West Indian manatee**
This is the largest living species of sirenian. Its streamlined shape enables it to swim through water with ease.

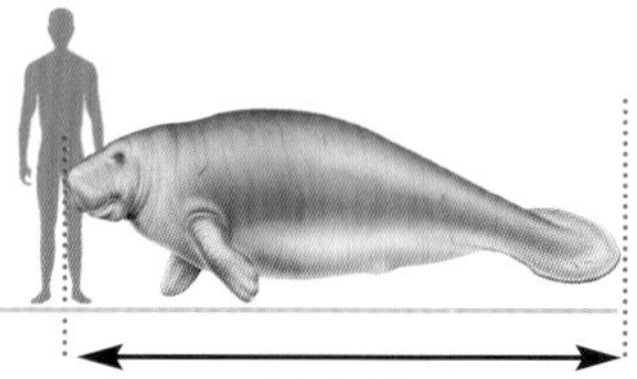

There are whiskers on the upper lip of the **snout**, *which is deeply cleft.*

nails

The relatively short **flippers** *are used to push food into the mouth. In the West Indian and West African manatee each flipper has four nails, which are absent in the Amazonian manatee.*

Sirenians have no hind limbs.

12–15 feet
(3.5–4.5 m)

arm enclosed in the animal's flank. The flippers serve a number of other uses. Manatees use their flippers to "walk" in shallow water along a seabed or riverbed. They also scratch themselves with their flippers, embrace other individuals, and direct food into the mouth. The flippers of West Indian and West African manatees have rudimentary nails at the tips of the concealed second, third, and fourth digits ("fingers"); the Amazonian manatee and the dugong do not have this feature. The flippers of dugongs are less flexible than those of manatees. As in whales, there are no hind limbs.

Manatee skin is wrinkled and tough, and it is up to 2 inches (5 cm) thick. Over most of the body it is only very sparsely haired. Around the lips are touch-sensitive, bristly hairs (vibrissae), which probably help the manatee to navigate in murky water and investigate the texture of its food quickly. Skin color ranges from slate gray to brown, often with an overgrowth of patches of green algae. Dugong skin is also gray or gray-brown, but it is less wrinkled than the manatee's and has shorter, more rigid hairs.

The sirenian head ends in a blunt muzzle with the pendulous upper lip hanging down over the mouth. The nostrils, located on the upper side of the muzzle, have valves that close when the animal dives.

Sirenian eyes are relatively small and lack well-defined eyelids, but they have a third, inner eyelid, called a nictitating membrane. This, along with heavy secretions of tears, cleans and protects the surface of the eyes. Sirenians see well underwater, but they rarely (if ever) use their eyes above the water. There are no external earflaps, and the ear canal opens onto the skin surface behind the eye.

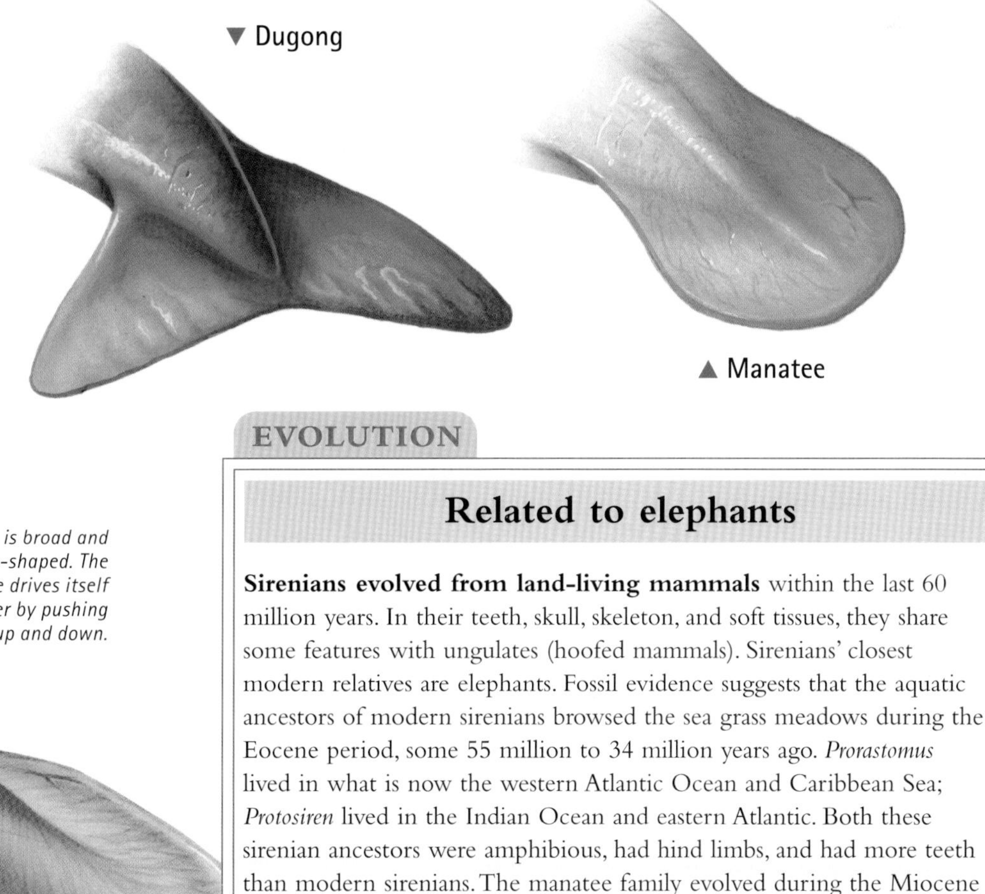

▼ **TAILS**
Sirenians
Sirenians move their tail up and down to swim. Manatee tails are large, broad, and paddle-shaped. Dugongs have a characteristic fluke-shape tail, like that of a whale.

EVOLUTION

Related to elephants

Sirenians evolved from land-living mammals within the last 60 million years. In their teeth, skull, skeleton, and soft tissues, they share some features with ungulates (hoofed mammals). Sirenians' closest modern relatives are elephants. Fossil evidence suggests that the aquatic ancestors of modern sirenians browsed the sea grass meadows during the Eocene period, some 55 million to 34 million years ago. *Prorastomus* lived in what is now the western Atlantic Ocean and Caribbean Sea; *Protosiren* lived in the Indian Ocean and eastern Atlantic. Both these sirenian ancestors were amphibious, had hind limbs, and had more teeth than modern sirenians. The manatee family evolved during the Miocene period (24 to 25 million years ago). The ancestors of today's manatees fed on reeds and grasses that grew in and around rivers, developing teeth that were replaced when the silica in the grass diet wore them down. In the past 100,000 years, the abundance and diversity of sirenian species have drastically declined.

▲ *Manatees (this is a West Indian manatee) and whales are only distantly related vertebrates, but in their use of forelimbs as flippers or fins, the loss of hind limbs, and the use of the tail as a paddle that moves up and down, they show convergent evolution. Manatees and whales have evolved similar anatomical solutions to the demands of moving in the same environment—water.*

Females have a pair of mammary glands, each with a nipple, in the "armpit" under the flipper. (Elephants' mammary glands are in a similar location, just inside the forelimb.) In males, the testes (sperm-producing organs) are contained within the abdomen, allowing greater streamlining—a feature found in other marine mammals, such as whales and seals.

In marine mammals, the fatty layer of blubber that lies beneath the skin is the main form of insulation that reduces loss of body heat into cold water. In living sirenians the blubber is relatively thin, compared with that of seals and whales of equivalent size. This, combined with their slow rate of metabolism and low heat output compared with more active marine mammals, probably restricts sirenians to warmer waters. Many sirenians live all year in tropical or subtropical waters. In temperate (mid-temperature) seas and rivers, sirenians migrate to warmer latitudes when water and air temperatures drop in the autumn.

COMPARATIVE ANATOMY

Snout and mouth

The profile of the snout and the position of the mouth in sirenian species vary according to their diet. The Amazonian and West African manatees eat mostly floating water plants, because the waters in which they feed are usually too murky for plants to grow beneath the surface. The snouts of these manatees do not slope down as markedly as those of other sirenian species. The dugong, whose diet consists largely of sea grasses that grow rooted in the seabed, has an underslung mouth. The West Indian manatee, with its varied diet of plants from the surface, middle, and bottom of the water, has a snout and mouth position between that of the dugong and the other manatees.

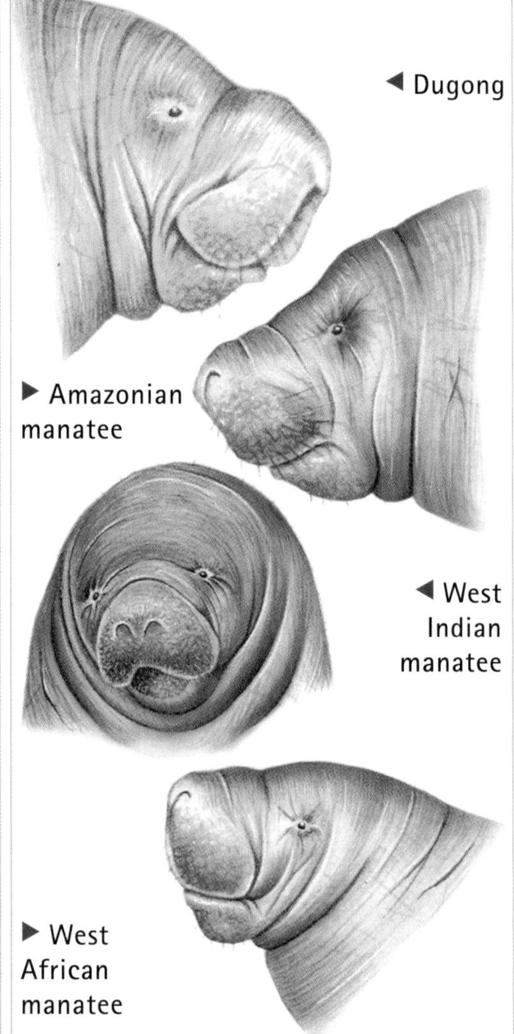

◀ Dugong

▶ Amazonian manatee

◀ West Indian manatee

▶ West African manatee

Skeletal system

In all species of vertebrates, the skeleton has several functions: it shapes and supports the animal's body; protects vital internal organs; and allows movement of body parts, thus enabling the animal to move around.

The skeleton of a sirenian is similar to that of a whale, but with some differences. Like whales, sirenians spend their entire life in water, which provides plentiful support, so they do not need strong limbs to support their body. Sirenians' use their forelimbs for steering through water rather than for support. Over millions of years of evolution, their hind limbs have shrunk and disappeared. Only the vestiges of a pelvis—which once connected the hind limbs and the spine—remain.

Skull and jaw

The skull and jaw of sirenians are heavily built, in part because of the need for firm attachment of the large muscles involved in grinding the large volumes of plant food they consume each day. In all sirenians, the forward part of the palate and the corresponding surface in the lower jaw are covered with rough, horny plates. These plates help in grasping and processing plant material before it is passed to molar teeth at the back for chewing.

In both manatees and dugongs, the nasal openings are set at or near the tip of the snout so that the animal does not have to raise much of its head above water to breathe. Manatees have nasal bones, but dugongs do not.

CONNECTIONS

COMPARE the teeth of a manatee with those of an *ELEPHANT*. These animals share a common ancestor within the last 100 million years, and both consume silica-rich grasses.

COMPARATIVE ANATOMY

Skulls of manatees and the dugong

The skull of an adult West Indian manatee averages 26 inches (66 cm) long. These animals typically have four to seven molar teeth at the back of each half jaw. New teeth erupt and move forward from the back to replace teeth toward the front that wear down and eventually fall out. The manatee is likely to lose 30 to 50 teeth in its lifetime.

The skull of an adult dugong averages 24 inches (61 cm) long. Its profile is more angular than that of the manatee, with the mouth more downward-pointing for feeding on bottom-living plants. There are only two or three molar teeth at the back of each half jaw. Adult male dugongs have forward-pointing incisor teeth that serve as tusks for ritualized fights with other males and for guiding or stimulating a mating partner. Growth rings form in cheek teeth enamel, and scientists use these to calculate the age of dugong specimens.

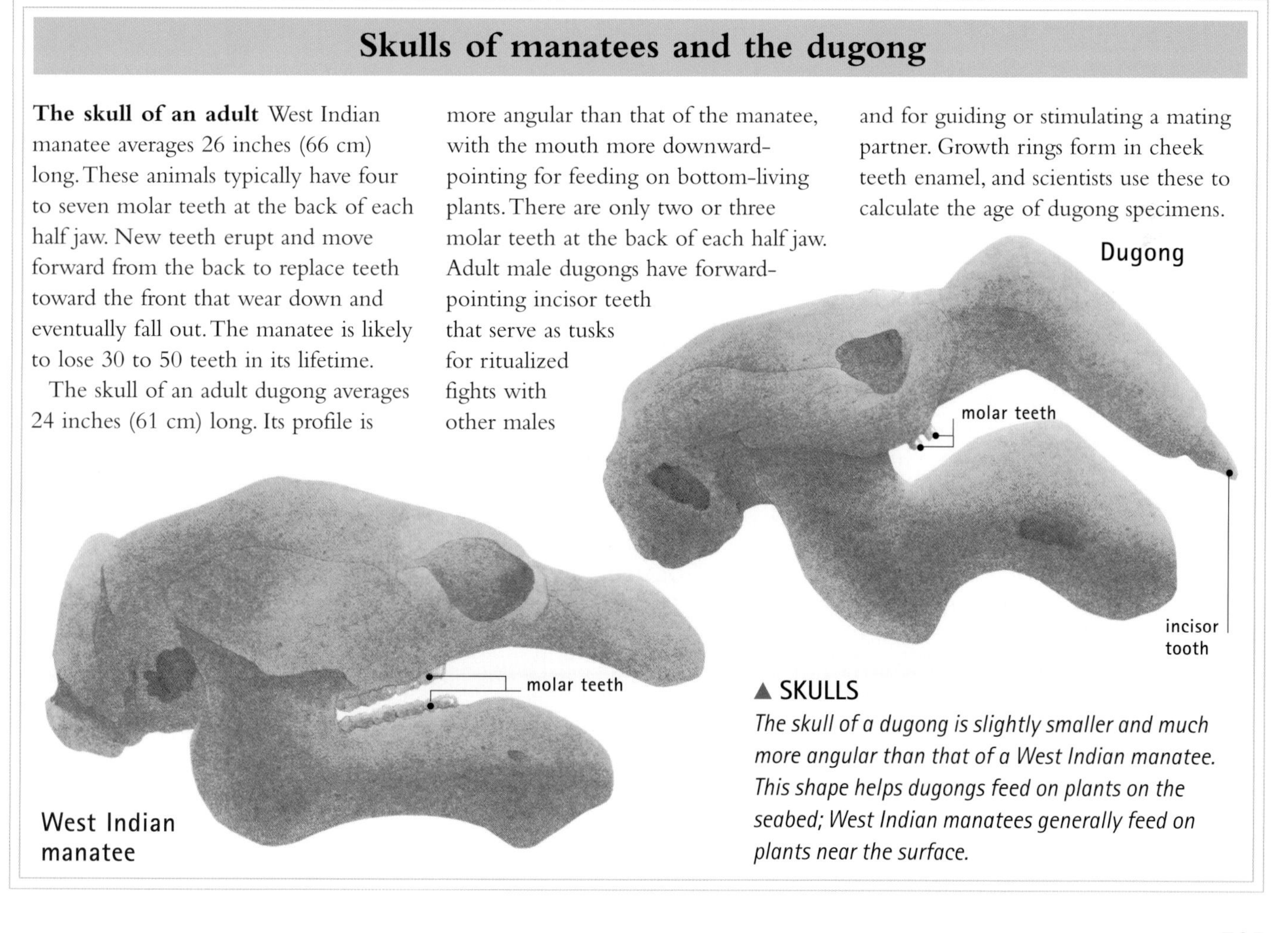

▲ SKULLS
The skull of a dugong is slightly smaller and much more angular than that of a West Indian manatee. This shape helps dugongs feed on plants on the seabed; West Indian manatees generally feed on plants near the surface.

COMPARATIVE ANATOMY

Sirenian skeletons

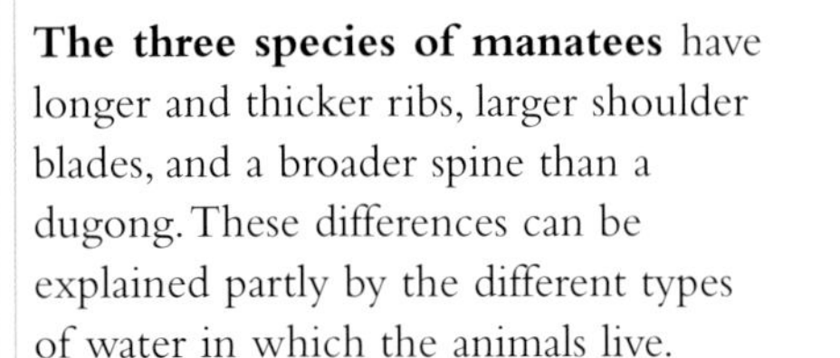

The three species of manatees have longer and thicker ribs, larger shoulder blades, and a broader spine than a dugong. These differences can be explained partly by the different types of water in which the animals live.

Dugongs live in seawater, and manatees spend some or all of their time in freshwater or brackish water. Freshwater and brackish water provide less buoyancy than seawater, so dugongs can rely more on the water to support their body.

▼ *The bones of all sirenians are very dense, but those of manatees are generally thicker and heavier than those of dugongs. The ancestors of sirenians were four-legged terrestrial animals, but none now has hind limbs.*

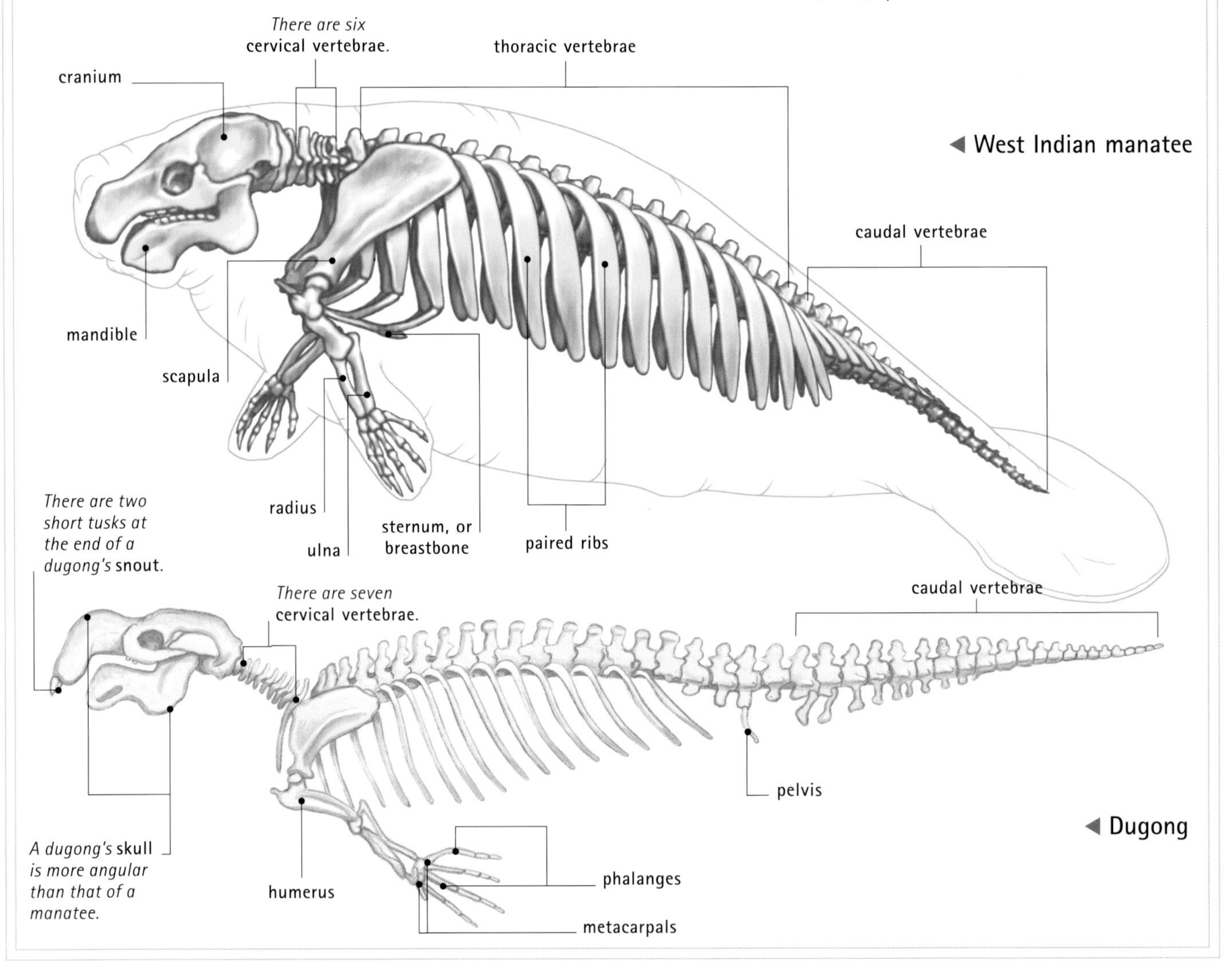

The vertebrae are separate and distinct throughout the sirenian spinal column. Manatees, in contrast to dugongs and almost all other mammals, have six cervical (neck) vertebrae instead of seven. In land mammals, the spine acts as a firm girder to support the animals' weight in air. Water helps support a sirenian's weight so the backbone is relatively more important for locomotion than it is for support. The number of vertebrae varies between individual manatees, but typically there are 56: 6 cervical (neck), 17 to 19 thoracic (chest), 3 sacral (originally concerned with attachment to the pelvis), and 23 to 29 caudal (tail). The neck is short, and the 6 cervical vertebrae are flattened along the axis of the spine. The neck moves little, so it does not increase drag or instability of the front end of the body during swimming.

The 17 to 19 ribs that attach to the thoracic vertebrae are very dense and probably serve as ballast: they help weigh the animal down in water. Behind the ribs, the vertebrae, especially those involved in moving the tail, have large transverse processes, or projections. Muscles attach to the processes. Beneath seven, eight, or nine of the tail vertebrae lie chevron-shape bones that help protect blood vessels from damage when the tail flexes. Strong, elastic ligaments running between the tail vertebrae allow the tail to bend and cushion and support the vertebrae, minimizing wear on the bones.

The last few caudal vertebrae are simple, flattened bones that support the tail at the center. The tail's paddle blades are supported by fibrous material in an arrangement similar to that found in the tail flukes of whales.

CLOSE-UP

Heavy bones

In contrast to whales, which have light bones, sirenians have heavy bones. The cavities in their long bones and ribs are almost entirely filled with hard bony tissue rather than fat-rich or blood-rich marrow, as in whales. In sirenians, heavy ribs probably act as ballast, helping to compensate for the high position of the air-filled lungs, which would otherwise make it difficult for the animal to twist and turn on its sides and back in a controlled way. Sirenians, with their plant diet, also produce considerable volumes of the gases methane and carbon dioxide in their intestines, which add to their buoyancy.

Limbs and their supports

In the terrestrial mammals from which sirenians evolved, the limbs were connected to the spine through limb girdles. The front girdle of land mammals typically contains two scapulae (shoulder blades) and clavicles (collar-bones). Over millions of years, the front limbs of the ancestors of today's sirenians evolved to become flippers.

Compared with a human arm, the skeleton of a sirenian's forelimb is similar but with the upper and lower arm bones shortened and the digits lengthened. The "fingers" are enclosed in skin and connective tissue that makes the outline of the flipper relatively smooth. The five digits characteristic of most land vertebrates are present but are not visible through the body surface. The presence of nails on the second, third, and fourth digits of West Indian and West African manatees hints at the pentadactyl (five-finger) structure that lies beneath. The flipper has the standard mammalian complement of phalanges, or finger bones, with the digits increasing in length from first to fifth to create the long, blunt-ended paddle shape.

In land mammals, the rear girdle (the pelvic girdle) is anchored to the backbone by sacral vertebrae that are fused together. During the sirenians' evolutionary transition over millions of years from a terrestrial to an aquatic mammal, the hind limbs have disappeared, along with most parts of the pelvis. Only small fragments of pelvic bone remain, and they provide points of attachment for muscles.

◀ *Although manatees are relatively slow-moving animals, their flippers, body shape, and buoyancy mechanisms make maneuvering in water easy. Here, a manatee enjoys an underwater back scratch.*

Muscular system

CONNECTIONS

COMPARE the muscles of a manatee's tail with those of another marine mammal, such as a ***DOLPHIN*** or a ***GRAY WHALE***. All have strong muscles, unlike the tail muscles of an ***ELEPHANT*** or a ***GRIZZLY BEAR***.

The manatee's muscle arrangement is suited for relatively slow movement, especially in brackish or freshwater. However, the dugong's tail muscles and whalelike tail blades enable rapid acceleration and fast swimming over short distances in coastal waters.

In manatees, a sheet of muscle extends from the pelvic region to the head. Over most of this area the muscle is more than 1 inch (2.5 cm) thick. Some muscle fibers and tendons extend into the forelimb; others pass beneath the neck. This massive muscle helps protect the organs inside the abdomen, substituting for the protective cartilage that extends from the sternum in other animals. When the muscle sheet on both sides of the tail contracts, together with abdominal muscles, this causes the tail to bend downward, producing the power stroke in forward propulsion. Contraction of muscle groups on the upper part of one side of the tail and relaxation of muscle groups on the opposite side cause the tail to tilt so that it acts as a rudder for steering.

The manatee uses its flippers in a variety of ways. They can be drawn forward to guide strands of weed toward the mouth. When the manatee is swimming fast, as when a male pursues a receptive female or escapes a predator, the flippers are held against the side of the body to minimize drag. The flippers can be extended slightly to help guide turns. When turned broadside against the direction of travel, the flippers act as brakes.

The flippers come into their own when the manatee is moving slowly or is stationary. On a seabed or riverbed, the flippers can be moved forward alternately so that the manatee "walks" along the bottom. When alarmed, the manatee pushes itself along the bottom using both flippers in unison. For slow swimming, the flippers are used like the oars or paddles of a rowboat. To turn to the right, the manatee pushes backward with its right flipper while pulling forward with its left. In "sculling" at low speed, the flippers are swung forward together and positioned to minimize resistance as they cut through the water. They are turned broadside for the backstroke, and are then tucked against the body before the next stroke. Manatees swim backward by moving the flippers in the reverse direction, without employing the tail.

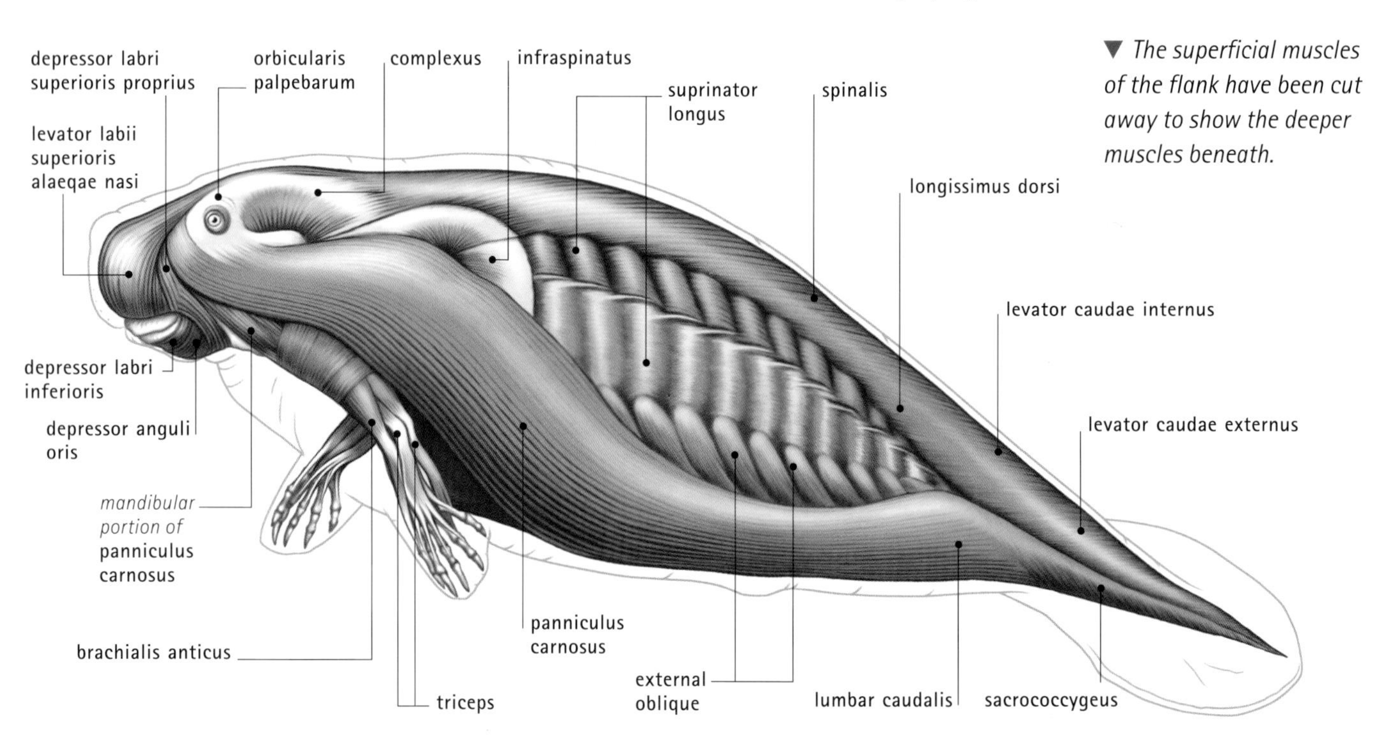

▼ *The superficial muscles of the flank have been cut away to show the deeper muscles beneath.*

Nervous system

As with other vertebrates, there are two main parts to a manatee's nervous system: the central nervous system, including the brain, and the peripheral nervous system. Compared to other marine mammals of similar size, sirenians have a small brain. The surface of their brain is also surprisingly smooth, with few of the surface folds normally associated with mammals of relatively high intelligence. The brain of an adult West Indian manatee weighs about 13 ounces (370 grams) on average, and that of a dugong about 11 ounces (300 grams). These weights are considerably less than the brain weights for a dolphin or sea lion of equivalent body size. Trainers are unable to teach captive sirenians to perform tasks, whereas they have had considerable success with sea lions, dolphins, and killer whales.

Brain structure

Recent studies of the manatee brain have shed light on aspects of manatee behavior and the relative importance of different sensory systems. The structures found in a manatee's brain are the same as those in most other mammals, but in manatees the relative importance of different regions is distinctive. The various parts of the midbrain and forebrain that relay and process sensory information from the mouth region, the flippers, and the tail are enlarged. The sense of touch (and possibly the detection of vibrations and water movement) is thus probably particularly significant in those parts of the body. In the hindbrain, the parts of the medulla concerned with moving the snout, lips, and sensory hairs around the mouth are also large. In the brainstem, relay centers concerned with contracting muscles of the tail and the diaphragm are large. The cerebellum, the part of the hindbrain that coordinates movement and balance, is large in sirenians, as would be expected for animals that spend their lives swimming.

The parts of the brain that are linked with social and emotional expression (such as the amygdala, the basal forebrain nuclei, and the hypothalamus) are not particularly large or elaborate in the manatee. This correlates well with the docile disposition of sirenians, with their apparent lack of complex social behavior, and with the absence of tight-knit social groups among them.

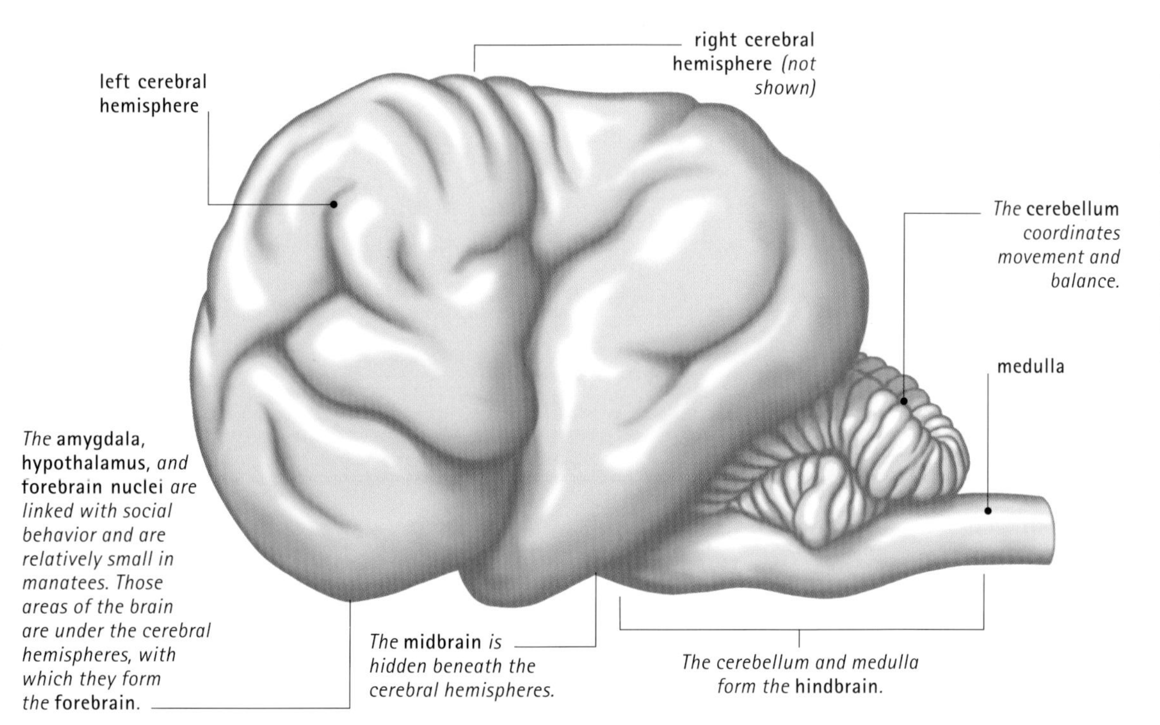

◀ **BRAIN West Indian manatee**
The surface of the cerebral hemispheres is relatively smooth, lacking the folds usually associated with high intelligence.

▲ *Despite having small eyes and no external earflaps, manatees see and hear well underwater.*

Sirenian senses

Sirenian eyes are quite small, but their behavior in the wild and in captivity suggests that they can see for long distances underwater. Manatee eyes see in color and have a reflective layer—the tapetum lucidum—behind the light-sensitive retina. This bounces light back through the retina, to maximize the light sensitivity of the eye. The tapetum lucidum is a feature of many birds and mammals that hunt at night or in poor light. It allows the manatee to see in murky water. At night, the tapetum lucidum gives manatee eyes a pinkish reflective shine like cats' eyes. Dugong eyes lack the tapetum lucidum and presumably can see less well in poor light.

Analysis of the structure of the parts of the brain and the nerve pathways concerned with vision suggests that in manatees this sense is not as highly developed as hearing and touch. The optic nerve and those parts of the brain that relay and process visual information are relatively small. The brain areas that control eye movement are also relatively small.

Manatees and dugongs have relatively good hearing from low to high frequencies (that is, pitches). This is borne out by the relatively large size of the parts of the hindbrain and midbrain that relay auditory, or sound, information, and of those parts of the cerebral cortex that interpret it.

Means of communication

Manatees produce sounds with the larynx (voice box), and they communicate with each other with middle- and low-frequency chirps and squeaks, and higher-frequency whistles, screams, and squeals. The precise meaning of these vocalizations is unclear, although they change according to whether or not the individuals are sexually aroused, frightened, or playing. Mothers and calves exchange chirps with one another and utter screams of alarm at times of danger.

Manatees also generate infrasonic sounds (frequencies too low for human hearing). These signals may be used by sexually active females to attract males. Some scientists speculate that manatees "echolocate" using sound—that is, produce sound beams or pulses that bounce off objects in their surroundings and are reflected back to them. This ability to "see with sound," which is well demonstrated in toothed whales, would be a valuable asset in the often murky waters where manatees live. However, sirenians do not appear to produce the pulsed and directed beams of sound that would be necessary for echolocation.

Manatees can probably smell only moderately well in water and perhaps in air. The olfactory nerves running from the organs of smell, and the parts of the brain that relay and interpret messages from these nerves, are only of modest size. The sense of taste is relatively well developed, as indicated by taste buds at the back of the tongue and prominent gustatory (taste) nerves running between them and the brain.

In manatees and dugongs, hairs scattered over the body sense water movement and direct contact with other objects, including other individuals. At certain times, such as when resting after feeding, manatees will interact with other members of the species and may act "playfully," rubbing against each other, gently grasping each other with their flippers, and even "kissing" snout to snout.

Circulatory and respiratory systems

As is typical for mammals, a sirenian has a four-chamber heart that pumps blood through a double circulation (the main and pulmonary circulations). Arteries with thick, muscular walls carry blood under high pressure away from the heart to supply other organs. Thin-walled veins carry blood back to the heart under low pressure. In the pulmonary (lung) circulation, carbon dioxide is expelled, and the blood is recharged with oxygen.

Like other mammals, sirenians inhale air through nasal passages, warming it before it travels down the trachea to the lungs. In the lungs, oxygen is exchanged for carbon dioxide, and the breath is exhaled when the animal surfaces. The respiratory system of a sirenian is more efficient than that of most terrestrial mammals, including humans.

Manatee and dugong lungs are unique in being positioned alongside the stomach and intestines, rather than in front of them. The manatee's chest cavity extends almost the entire length of the trunk, and this probably produces great benefits in terms of buoyancy control. The manatee's lungs are also unusual in having fewer bronchi (primary air passages) and fewer major blood vessels than are found in the lungs of other mammals. The dugong too has simple lungs, each with a main bronchus that runs almost the entire length of the lung with only a few side branches. Manatee lungs contain large amounts of smooth muscle and elastic fibers, which enable the animal to fine-tune lung volume to adjust buoyancy during a dive.

IN FOCUS

Adjusting buoyancy

Manatees have an unusual mechanism for adjusting their buoyancy. Their lungs are unusually elongated and extend alongside the gut, rather than in front of the gut as in other mammals, including whales. Thus the lungs help the animal float in a horizontal position. The lungs, being relatively narrow, are also less affected by pressure differences between the top and bottom of the lung during diving. In the manatee, the diaphragm—the sheet of muscle separating the thoracic (chest) cavity from the abdominal cavity—is also unusually large. By contracting or relaxing the diaphragm, the animal can reduce or increase the volume of air in the lungs, thus adjusting overall buoyancy. For example, when the diaphragm contracts, the lung volume reduces, and the animal sinks in the water. Relaxing the diaphragm causes the manatee to rise. This mechanism is almost effortless, and is a more efficient version of the way human scuba divers learn to adjust their buoyancy by breathing in or out.

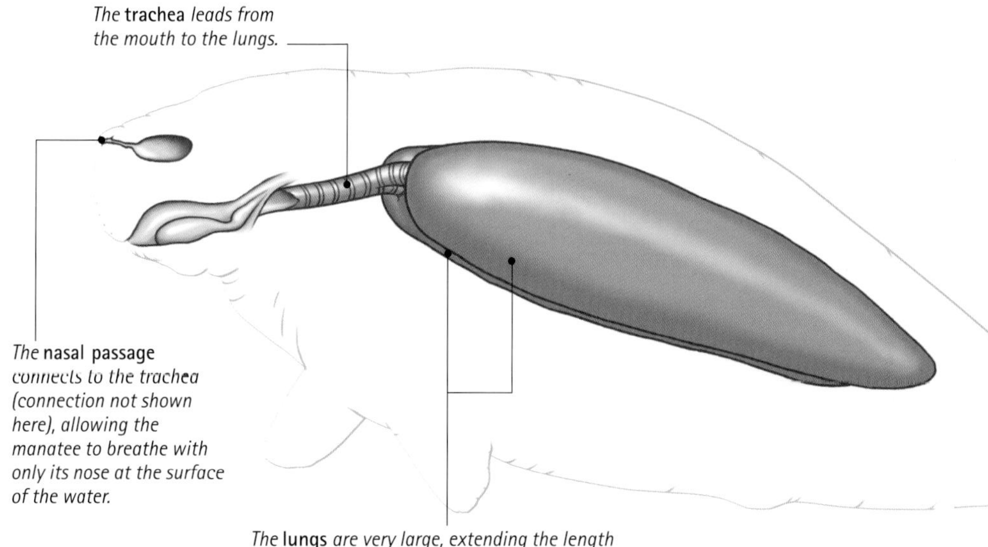

◀ RESPIRATORY SYSTEM

A manatee's respiratory system is suited to an air-breathing aquatic lifestyle. The large lung capacity provides considerable buoyancy.

▲ *Although they do not dive deep, manatees are well adapted for plunging underwater. Their lungs inflate with inhaled air very rapidly, and their nostrils have valves that close when the animal dives.*

Unlike whales and seals, sirenians do not dramatically slow their heart rate when they dive. The Amazonian manatee's heart, for example, beats about 50 times a minute when the animal is at the water surface and slows slightly to about 30 to 40 beats a minute during normal dives. Only when a manatee is threatened does its heart rate plummet to as low as five or six beats a minute—a change as dramatic as that found in deep-diving whales and seals—allowing it to remain safely submerged for a longer period.

Manatees, like other marine mammals, cannot breathe underwater, so they store the air they require for the dive in their respiratory system. Oxygen is also stored temporarily in hemoglobin and myoglobin, the oxygen-carrying pigments in blood and muscle. Manatee blood does not have the high levels of hemoglobin that is found in active, deep-diving marine mammals such as most whales and seals. This is understandable, as manatees do not usually dive as deep or for as long, and are much less active, and so do not require large amounts of oxygen to provide the energy for muscle contraction. This lower demand for oxygen relative to other sea mammals is also indicated by the low levels of myoglobin. In whales and seals, this substance is present at high levels in the muscles and gradually releases oxygen during dives. In manatees, myoglobin (as well as hemoglobin) levels are similar to those found in land mammals.

IN FOCUS

Lungs for diving

Sirenians dive with their lungs inflated. In this regard they are similar to whales but different from seals, which hold minimal air in their lungs during a deep dive. Manatees can stay submerged for 20 minutes when resting or swimming slowly, and dugongs can stay underwater for about half this time. Manatee lungs are efficient, and—like the lungs of whales—can exchange nearly 90 percent of the air in them with each breath; the figure for humans is typically only 15 to 20 percent.

Digestive and excretory systems

Sirenians feed on water plants, and exclude most of the water when they swallow. The West Indian manatee eats up to about 40 species of aquatic plants, including floating water hyacinth and rooted sea grasses, as well as 10 kinds of algae. The dugong, with its more downturned snout, prefers sea grasses. Sirenians will also eat attached or slow-moving marine invertebrates such as sea squirts and sea cucumbers. Manatees have been found eating fish caught in gill nets, but for these animals fish is not a usual part of the diet.

Most mammals are "diphyodont"—that is, they have milk teeth when juvenile, which are later replaced with one set of permanent teeth In contrast, tooth replacement in manatees is almost continuous (polyphyodont) throughout their lives. Replacement of the first teeth is triggered when the young calf begins to add vegetable matter to its milk diet.

The dugong has an unusual arrangement of lips and mouthparts for consuming marine grasses and their rootlike rhizomes. The upper lip is extended to form a heavily bristled muscular pad with a deep cleft, which overhangs the downward-pointing mouth. The dugong grasps the base of a sea grass plant with one or both sides of the muscular pad. Horny pads toward the front of the dugong's jaws help pull the plant out of the sediment along with its roots or rhizomes (which are rich in carbohydrates) or, in the case of larger sea grasses, break off the stem. This feeding action is similar to the way in which cattle use their tongue and lips to grasp grass stems. The sea grass material is passed to the back of the mouth to be ground up by molar teeth. The dugong is so efficient at grazing small sea grasses that it leaves bare seabed.

The West Indian manatee has an even more adaptable upper lip arrangement than the dugong. The upper lip is less deeply divided than the dugong's, and is used in feeding at the surface. The lower lip is used for taking food that is growing lower in the water. Like the dugong, this manatee can wrap its top lip around objects and pluck them up, with an action almost like big soft tweezers.

Sirenians, like land-living plant-eaters, produce saliva from glands in the mouth to lubricate their food and begin the process of digestion. Once ingested, food passes down a muscular esophagus into a two-part stomach through a strong ring of muscle, the cardiac sphincter, which acts as a valve. The first chamber, or main stomach, has thick, muscular

CONNECTIONS

COMPARE the muscular top lip of a sirenian with the trunk (proboscis) of an ***ELEPHANT***. Both are effective solutions for grasping and uprooting plant material without the use of limbs.

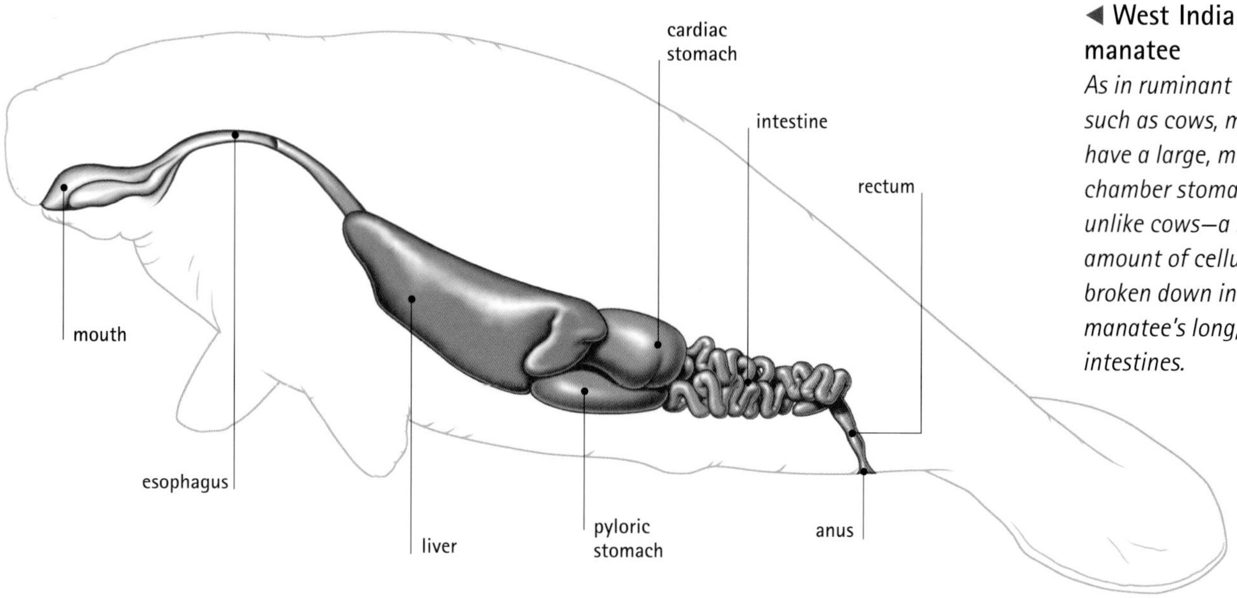

◀ West Indian manatee
As in ruminant animals such as cows, manatees have a large, multi-chamber stomach, but—unlike cows—a large amount of cellulose is broken down in the manatee's long, coiled intestines.

▲ *Dugongs need to eat large amounts of sea grass. Typically, they feed at depths of 3 to 16 feet (1 to 5 m) and stay underwater for between one and three minutes. Dugongs shake their head during feeding, apparently to clean sediment from the food before it is swallowed. Very little sediment has been found in the stomach contents of dugongs that scientists have dissected.*

walls that churn the contents and release digestive enzymes. These are added to by secretions from two outpockets of the stomach called gastric ceca. The second stomach chamber is smaller than the first, with thinner walls, and it quickly pushes its food contents into the small intestine.

Sirenians are nonruminants, like horses and elephants. Although they have a stomach with compartments, they do not use it—as ruminants such as cows and sheep do—to digest plant material slowly by fermentation. Instead, similar processes take place in the hind part of the gut, the intestines. Sirenian intestines are extremely long and thus able to break down the low-quality vegetation on which they feed. In a large manatee, the small intestine is some 50 feet (15 m) long, and the large intestine is of similar length. In a dugong, the large intestine is up to 100 feet (30 m) long. Bile from the liver and pancreatic juice from the pancreas empty into the first part of the small intestine. Bile adds salts that help break down fats into droplets, and pancreatic juice adds a range of digestive enzymes. Between the small and large intestine lies a side branch—a blind-ended sac called the cecum. This contains symbiotic (partnership) bacteria that digest cellulose, the complex carbohydrate found in the cell wall of plant cells, which is difficult to digest. Food takes about seven days to pass through the digestive system of a sirenian. In a day, sirenians eat 8 to 15 percent of their body weight, which in a large manatee can amount to more than 200 pounds (90 kg).

IN FOCUS

Slow metabolism

Sirenians do not expend much energy: their metabolic rate (the rate at which food and oxygen are used to release energy and heat) is less than one-third that of other marine mammals of similar weight.
This feature can be partly explained because sirenians live in a warm environment without rapid changes in temperature, so they do not need to use much energy in keeping warm or regulating their body temperature. In addition, they are quite sluggish and slow-moving, except on rare occasions—such as when they are threatened by large predators.

Gaining water and losing salt

In the wild manatees obtain freshwater from their diet, or they need to drink it from their surrounding water. This restricts them to living in freshwater or brackish water for at least part of the time, and consuming plants that are low in salts and high in water content. Probably, they can live only in full-strength seawater and consume salt-rich sea grasses for short periods. The kidneys of manatees have long loops of Henle, which suggest they can produce concentrated urine as a means of expelling excess salts and retaining valuable water at the same time.

Reproductive system

The reproductive system of sirenians is essentially the same as that of placental land mammals. However, there are some major differences due to the need for streamlining and for the animals to mate, give birth, and suckle under water. To maximize streamlining, male sirenians have an internal penis and testes. Most of the time, the penis lies inside the abdomen. Before mating, it fills with blood and emerges through the genital slit.

Mating

Usually, manatees form only loose associations, such as gathering in safe water during the winter. In warm weather, however, they become sexually active, and a dozen or more males will track a sexually receptive female in order to mate with her. They may follow her for several weeks, competing with each other to mate. Apparently an unwilling participant, she finally gives in to one or more of the most persistent males. A pair usually mate face-to-face with the male beneath the female, clasping her with his flippers. After mating, the males

IN FOCUS

Vulnerable

Manatees and the dugong have low reproductive rates. Often, even long-lived female dugongs aged 50 or more have produced fewer than 10 young. This fact, combined with encounters with humans in which sirenians are hunted or accidentally injured and the habitats in which they live are disturbed or destroyed, puts these animals in danger. The International Union for the Conservation of Nature and Natural Resources (IUCN) classes all sirenian species as vulnerable, and in many regions both local and international laws seek to protect them. One estimate suggests that, if more than 2 percent of adult females in a dugong population are hunted each year, the population will be unable to replace its losses and will go into grave decline.

▼ POSITION OF MALE AND FEMALE GENITALS
West Indian manatee

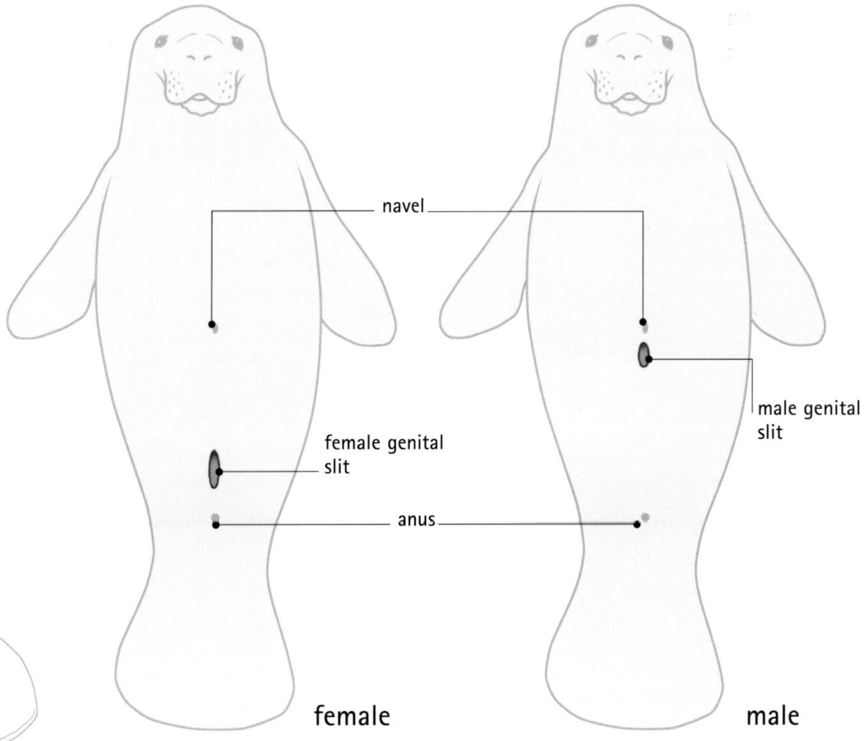

▼ FEMALE REPRODUCTIVE ORGANS
An egg is released from one of the ovaries and is fertilized with sperm from the male. The fertilized egg develops in the uterus, and the female gives birth through the vagina after a gestation of 12 or 13 months.

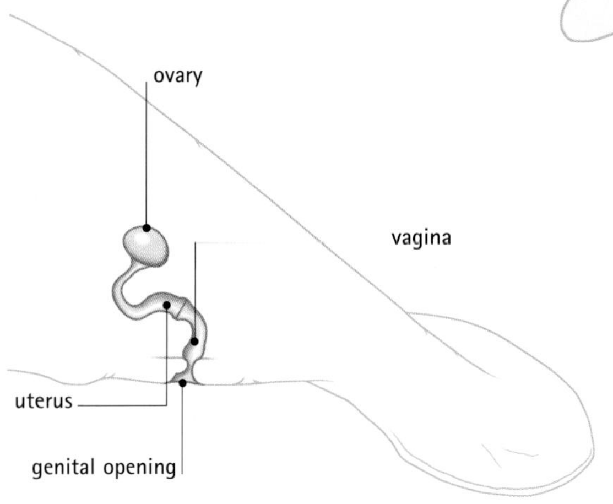

◀ *Manatees mate throughout the year, but mating occurs most commonly at a season when the water is warmest. This warmth allows the calf to be born, one year later, when food supplies for the suckling mother are at their peak.*

disperse to find other females, and the potential fathers play no further role in the life of mother or calf.

Dugong mating behavior is broadly similar to that of manatees, but competition between males is even more intense. Mature males will fight with each other to establish a territory and gain access to one or more females in the vicinity. These ritual fights, in which one male uses his tusks and body weight to gain advantage over the other, can leave males with deep scars and even more serious injuries. Before mating, a male also uses his tusks to help turn the female onto her back.

▲ **Mating behavior**
A dozen or more male manatees will pursue a receptive female, with one eventually mating with her.

IN FOCUS

Scent messages

Observers often see manatees rubbing their genitalia, armpits, and head—regions where glandular secretions are released—against particular objects in the water, such as rocks or logs. An individual will use the same "rubbing post" year after year. In all likelihood, the animal is scent-marking. Leaving scent can provide a variety of informative messages about the individual's condition, including its readiness to mate.

▼ **Social bond**
Mother and calf form a strong social bond. They greet each other snout to snout, and communicate using a variety of calls.

IN FOCUS

The cradling myth and mermaids

The presence of the sirenian's two mammary glands in the chest region close to her flippers has given rise to fanciful tales. One describes the mother cradling the calf "in her arms" while it suckles. In fact, the calf usually suckles while the mother swims upright—although the mother does sometimes swim on her back with the calf on top when they play together. The arrangement of the mother's flippers and mammary glands has led to sailors to imagine the upper half of the sirenian's body as that of a woman, with the lower half having a tail like that of a fish (although it is more like that of a whale or giant otter). This is probably how the mermaid legend began.

▲ Birth
Female manatees almost always give birth to just one calf per pregnancy. The calf is able to swim as soon as it is born.

▼ Suckling
A manatee calf swims alongside its mother to suckle from the nipple behind the base of the flipper.

Pregnancy and birth

Manatees do not become mature until they are six to eight years old, and dugongs not until their teens. Females typically produce only one calf at a time, and they often have intervals of several years between calves. In sirenians, the gestation period (the time from copulation to birth) is 12 to 13 months, similar to that of many whales. Often breeding is timed to occur in the warmer months, so that the calf is born the following year in the season when vegetation is abundant and provides plenty of nutrition for the suckling mother.

Manatees and the dugong usually give birth to one calf, which is often born tail first, like most toothed whales. This may be an adaptation to delay the time when the calf has to take its first breath, until the last possible moment. As a fetus, the manatee is covered in fine downy hair, but at birth it is more or less hairless. A newborn West African or West Indian averages about 4 feet (1.2 m) long and weighs 65 pounds (about 30 kg). Dugong and Amazonian manatee calves are slightly smaller. A newborn calf can usually swim unaided, but its mother will often help it to the surface to breathe.

Suckling and weaning

The manatee calf suckles its mother's milk for 12 to 18 months; the dugong for up to 24 months. In both cases, however, the calves also graze vegetation when only a few months old. The mother's mammary glands are in the axillary, or armpit, position close to the base of the flipper. The calf swims slightly below and to one side of her to take her milk, which is rich in fats, proteins, and salts. The mammary glands lack storage sacs, as found in cattle and goats, and so the calf suckles for a short time at regular intervals. Mother and calf have a strong social bond and often greet each other with snout-to-snout nuzzling that looks like kissing.

TREVOR DAY

Further Reading and Research

Perrin, W. F., B. Würsig, and J. G. M. Thewissen (eds.). 2002. *Encyclopedia of Marine Mammals.* Academic: San Diego, CA.

Reynolds, John E., III, and S. A. Rommel (eds.). 1999. *Biology of Marine Mammals.* Smithsonian Institution Press: Washington, D.C.

Index

Page numbers in **bold** refer to main articles; those in *italics* refer to picture captions, or to names that occur in family trees.